Advances in Intelligent Systems and Computing

Volume 651

Series editor

Janusz Kacprzyk, Polish Academy of Sciences, Warsaw, Poland
e-mail: kacprzyk@ibspan.waw.pl

About this Series

The series "Advances in Intelligent Systems and Computing" contains publications on theory, applications, and design methods of Intelligent Systems and Intelligent Computing. Virtually all disciplines such as engineering, natural sciences, computer and information science, ICT, economics, business, e-commerce, environment, healthcare, life science are covered. The list of topics spans all the areas of modern intelligent systems and computing.

The publications within "Advances in Intelligent Systems and Computing" are primarily textbooks and proceedings of important conferences, symposia and congresses. They cover significant recent developments in the field, both of a foundational and applicable character. An important characteristic feature of the series is the short publication time and world-wide distribution. This permits a rapid and broad dissemination of research results.

More information about this series at http://www.springer.com/series/11156

Shabana Urooj · Jitendra Virmani
Editors

Sensors and Image Processing

Proceedings of CSI 2015

Springer

Editors
Shabana Urooj
Gautam Buddha University
Greater Noida, Uttar Pradesh
India

Jitendra Virmani
Department of Electrical and
Instrumentation Engineering
Thapar University
Patiala, Punjab
India

ISSN 2194-5357 ISSN 2194-5365 (electronic)
Advances in Intelligent Systems and Computing
ISBN 978-981-10-6613-9 ISBN 978-981-10-6614-6 (eBook)
https://doi.org/10.1007/978-981-10-6614-6

Library of Congress Control Number: 2017952512

Printed on acid-free paper

This Springer imprint is published by Springer Nature
The registered company is Springer Nature Singapore Pte Ltd.
The registered company address is: 152 Beach Road, #21-01/04 Gateway East, Singapore 189721, Singapore

Preface

The last decade has witnessed remarkable changes in IT industry, virtually in all domains. The 50th Annual Convention, CSI-2015, on the theme "Digital Life" was organized as a part of CSI-2015, by CSI at Delhi, the national capital of the country, during December 02–05, 2015. Its concept was formed with an objective to keep ICT community abreast of emerging paradigms in the areas of computing technologies and more importantly looking at its impact on the society.

Information and Communication Technology (ICT) comprises of three main components: infrastructure, services, and product. These components include the Internet, infrastructure-based/infrastructure-less wireless networks, mobile terminals, and other communication mediums. ICT is gaining popularity due to rapid growth in communication capabilities for real-time-based applications. "Nature Inspired Computing" is aimed at highlighting practical aspects of computational intelligence including robotics support for artificial immune systems. CSI-2015 attracted over 1500 papers from researchers and practitioners from academia, industry, and government agencies, from all over the world, thereby making the job of the Programme Committee extremely difficult. After a series of tough review exercises by a team of over 700 experts, 565 papers were accepted for presentation in CSI-2015 during the 3 days of the convention under ten parallel tracks. The Programme Committee, in consultation with Springer, the world's largest publisher of scientific documents, decided to publish the proceedings of the presented papers, after the convention, in ten topical volumes, under ASIC series of the Springer, as detailed hereunder:

1. Volume # 1: ICT Based Innovations
2. Volume # 2: Next Generation Networks
3. Volume # 3: Nature Inspired Computing
4. Volume # 4: Speech and Language Processing for Human-Machine Communications
5. Volume # 5: Sensors and Image Processing
6. Volume # 6: Big Data Analytics

7. Volume # 7: Systems and Architecture
8. Volume # 8: Cyber Security
9. Volume # 9: Software Engineering
10. Volume # 10: Silicon Photonics and High Performance Computing

We are pleased to present before you the proceedings of Volume # 5 on "Sensors and Image Processing." The title "Sensors and Image Processing" highlights the different applications in the field of virtual reality. It also delves into the matter as to how robotics can be applied to strengthen modeling. The title also showcases the various augments of latest mobile technologies, solid modeling, etc.

Sensors are used in everyday objects such as touch-sensitive elevator buttons and lamps which dim or brighten by touching the base, besides innumerable applications of which most people are never aware of. With advances in micro-machinery and easy-to-use microcontroller platforms, the uses of sensors have expanded beyond the more traditional fields of temperature, pressure, or flow measurement, for example MARG sensors. Image processing is processing of images using mathematical operations by using any form of signal processing for which the input is an image, such as a photograph or video frame. The output of image processing may be either an image or a set of characteristics or parameters related to the image. Most image-processing techniques involve treating the image as a two-dimensional signal and applying standard signal-processing techniques to it. The title "Sensors and Image Processing" also amalgamates and showcases the applications of above technologies in different research and real-time domains. The volume includes scientific, original, and high-quality papers presenting novel research, ideas, and explorations of new vistas in speech and language processing such as speech recognition, text recognition, embedded platform for information retrieval, segmentation, filtering and classification of data, emotion recognition. The aim of this volume is to provide a stimulating forum for sharing knowledge and results in the model, methodology, and implementations of speech and language processing tools. Its authors are researchers and experts of these domains. This volume is designed to bring together researchers and practitioners from academia and industry to focus on extending the understanding and establishing new collaborations in these areas. It is the outcome of the hard work of the editorial team, who have relentlessly worked with the authors and steered up the same to compile this volume. It will be a useful source of reference for the future researchers in this domain. Under the CSI-2015 umbrella, we received over 200 papers for this volume, out of which 29 papers are being published, after rigorous review processes, carried out in multiple cycles.

On behalf of organizing team, it is a matter of great pleasure that CSI-2015 has received an overwhelming response from various professionals across the country. The organizers of CSI-2015 are thankful to the members of *Advisory Committee, Programme Committee, and Organizing Committee* for their all-round guidance, encouragement, and continuous support. We express our sincere gratitude to the learned *Keynote Speakers* for support and help extended to make this event a grand success. Our sincere thanks are also due to our *Review Committee Members* and the

Editorial Board for their untiring efforts in reviewing the manuscripts, giving suggestions and valuable inputs for shaping this volume. We hope that all the participated delegates will be benefitted academically and wish them for their future endeavors.

We also take the opportunity to thank the entire team from Springer, who have worked tirelessly and made the publication of the volume a reality. Last but not least, we thank the team from Bharati Vidyapeeth's Institute of Computer Applications and Management (BVICAM), New Delhi, for their untiring support, without which the compilation of this huge volume would not have been possible.

Greater Noida, India — Shabana Urooj
Patiala, India — Jitendra Virmani
March 2017

The Organization of CSI-2015

Chief Patron

Padmashree Dr. R. Chidambaram, *Principal Scientific Advisor, Government of India*

Patrons

Prof. S.V. Raghavan, Department of Computer Science, IIT Madras, Chennai
Prof. Ashutosh Sharma, Secretary, Department of Science and Technology, Ministry of Science and Technology, Government of India

Chair, Programme Committee
Prof. K.K. Aggarwal, Founder Vice Chancellor, GGSIP University, New Delhi

Secretary, Programme Committee
Prof. M.N. Hoda, Director, Bharati Vidyapeeth's Institute of Computer Applications and Management (BVICAM), New Delhi

Advisory Committee

- Padma Bhushan Dr. F.C. Kohli, Co-Founder, TCS
- Mr. Ravindra Nath, CMD, National Small Industries Corporation, New Delhi
- Dr. Omkar Rai, Director General, Software Technological Parks of India (STPI), New Delhi
- Adv. Pavan Duggal, Noted Cyber Law Advocate, Supreme Courts of India
- Prof. Bipin Mehta, President, CSI
- Prof. Anirban Basu, Vice President—cum- President Elect, CSI
- Shri Sanjay Mohapatra, Secretary, CSI
- Prof. Yogesh Singh, Vice Chancellor, Delhi Technological University, Delhi
- Prof. S.K. Gupta, Department of Computer Science and Engineering, IIT, Delhi

- Prof. P.B. Sharma, Founder Vice Chancellor, Delhi Technological University, Delhi
- Mr. Prakash Kumar, IAS, Chief Executive Officer, Goods and Services Tax Network (GSTN)
- Mr. R.S. Mani, Group Head, National Knowledge Networks (NKN), NIC, Government of India, New Delhi

Editorial Board

Contents

About the Editors

Dr. Shabana Urooj currently works at Gautam Buddha University, Greater Noida, India. She received her bachelor's in Electrical Engineering and M.Tech. in Instrumentation and Control from Aligarh Muslim University, Aligarh, India. She has completed her doctorate at Jamia Millia Islamia (A Central University), New Delhi, India. She has 3 years of industrial and 15 years of teaching experience. She has authored/co-authored more than 90 research papers. She has served as an editor, board member, and reviewer for a number of leading journals. She is an active volunteer of IEEE and actively involved in various professional development ventures and bodies, such as the CSI and the ISTE.

Jitendra Virmani received his B.Tech. (Hons) in Instrumentation Engineering from Sant Longowal Institute of Engineering and Technology, Punjab, in 1999; his M.Tech. in Electrical Engineering with specialization in measurement and instrumentation from the Indian Institute of Technology Roorkee, India, in 2006; and his Ph.D. on Analysis and Classification of B-Mode Liver Ultrasound Images from the same institute in 2014. He held various academic posts before joining the Department of Electrical and Instrumentation Engineering, Thapar University, Patiala, Punjab, India, where he has been an Assistant Professor since July 2015. He is a life member of the Institute of Engineers (IEI), India. His research interests include application of machine learning and soft computing techniques for the analysis of medical images.

A Comparative Study of Various Color Texture Features for Skin Cancer Detection

Maurya Ritesh and Srivastava Ashwani

Abstract Detection of skin cancer gives the best chance of being diagnosed early. Biopsy method for skin cancer detection is much painful. Human interpretation contains difficulty and subjectivity; therefore, automated analysis of skin cancer-affected images has become important. This paper proposes an automatic medical image classification method to classify two major type skin cancers: melanoma and non-melanoma. In this paper, we have used the color and texture features in combination which gives better results than using color or gray-level information alone. We have used k-means clustering algorithm to segment the lesion. The features are extracted by seven different color texture feature extractors from the segmented images. Classification accuracy of our proposed system is evaluated on four different types of classifiers, and their values are compared with one another. The results of the proposed system are computed on fivefolds of cross-validation in order to perform better analysis of our proposed method.

Keywords Gray-level co-occurrence matrices · Support vector machine · Local binary patterns · Texture features · Color percentiles · K-means clustering · Co-occurrence matrix · Color features · Integrative co-occurrence matrix · Gabor features · Linear classifier · NN classifiers · NMC classifiers · Cross-validation

M. Ritesh (✉)
Department of Computer Science and Engineering, Shri Ramswaroop Memorial University, Lucknow, U.P, India
e-mail: maurya123ritesh47@gmail.com

S. Ashwani
Computer Engineering and Application Department, GLA University, Mathura, India
e-mail: ashwani.sri89@gmail.com

S. Urooj and J. Virmani (eds.), *Sensors and Image Processing*, Advances in Intelligent Systems and Computing 651, https://doi.org/10.1007/978-981-10-6614-6_1

1 Introduction

Skin cancer incidence is increasing at 3.1% per year [1]. Skin cancer spread over the body with the help of lymphatic and blood vessels. Thus, early detection of skin cancer is very important for proper diagnosis of the disease.

Melanoma and non-melanoma are two major categories of skin cancers. Malignant melanoma is of several subtypes. Basal cell carcinoma and squamous cell carcinomas are two main types of non-melanoma skin cancers.

Each type of skin cancer is different from the other skin cancers in certain characteristics.

In clinical detection of skin cancer diagnosis, dermatologists use a visual inspection. Clinical diagnostic performance is very poor in comparison with dermoscopy and automatic diagnosis. Dermoscopy is a noninvasive diagnostic technique. It uses clinical dermatology and dermatopathology in combination to inspect the morphological features which is not possible in clinical detection. Dermoscopy increases the performance of diagnosis with 10–30% compared to unaided eye [2]. Differentiation of skin cancer images needs much more experience with dermoscopy technique. Less-experienced clinicians use ABCD-E rule [3] to improve the diagnostic performance [3].

Automatic image processing of skin cancer gives better results by providing the exact information about lesion, which can be useful for the clinician to detect and classify skin cancer. It is also used as a stand-alone early warning tool. Effective implementation of this automatic technique may give reduced deaths with benefits both to the patients and to the healthcare system. Working of automatic skin cancer detection has three main stages: (1) segmentation of lesion; (2) feature extraction and feature selection; and (3) lesion classification.

Segmentation is an important process in image processing applications and computer vision because doctors are always interested in meaningful regions of the dermoscopy image. Segmentation divides an image into a number of separate regions. Pixels, in each region, have high similarity such as color, intensity, and texture. Many researchers use only gray level for image segmentation [31]. But, in our proposed system, we use color information of the image for lesion segmentation.

In general, we covert the color image in gray-level image; therefore, color information does not used. There is a wide variety of segmentation methods used in dermoscopy images [4]. Recent advancements include thresholding [5, 6], k-means clustering [7], fuzzy c-means clustering [8, 9], density-based clustering [10], mean-shift clustering [11], gradient vector flow snakes [12–14], color quantization followed by spatial segmentation [15], statistical region merging [16], watershed transformation [17], dynamic programming [18, 19], and supervised learning [20, 21]. Clustering is an unsupervised learning technique, where one can give the number of clusters in advance to classify pixels [22]. A similarity measure is defined between pixels, and similar pixels are then grouped into a cluster. We use k-means clustering for segmentation of color images.

It is very hard to differentiate skin cancer visually. Identification and extraction of most effective features from cancer-affected lesion is very important. Each class of skin cancer has some different features than others. We use these different features for classification. Feature extraction extracts useful features or properties from original dataset of an image. These extracted features easily classify the classes of skin cancer.

Color features are mainly statistical parameters. These are calculated from inter- and intra-channel of an image, like average value and standard deviation of the RGB [23–27] or HSV color channels [28]. Here, we use "local binary patterns + color percentiles," "integrative co-occurrence matrices," "gray-level co-occurrence matrices + color percentiles," "Gabor features + chromatic features," "Gabor features," "opponent color LBP," "color ranklets" [29]. These methods are based on texture and disjoint color analysis. Textural features are extracted from images by converting into gray level, and color features are computed with the help of three color component of an image. Textural and color features are concatenated into the same feature vector to improve the classification accuracy.

The main aim of feature selection is to select the maximum number of features to achieve high performance in cancer classification [30]. Feature selection is important when anyone works on gray-level features. In our proposed system, there is no need of feature selection algorithms.

Classification phase of the diagnostic system is the one in charge of making the inferences about the extracted information in the previous phases in order to be able to produce a diagnostic about the input image. In our experiment, we have used four well-established classifiers: support vector classifier (SVC), nearest neighborhood (NN), linear classifier, and nearest mean classifier (NMC).

The rest of the paper is organized as follows: Sect. 2 briefly reviews segmentation, feature extraction techniques, and classification methods which are used in proposed framework. Section 3 reports extensive experimental results, and Sect. 4 concludes the paper.

2 Proposed Framework

Proposed framework is a compiled abstraction of digital image classification. Figure 1 describes the steps in an image classification.

In this paper, we propose this framework with k-means clustering segmentation and color texture feature extraction techniques which is a new approach to classify the skin cancer images. For validation purpose, twofold cross-validation is used.

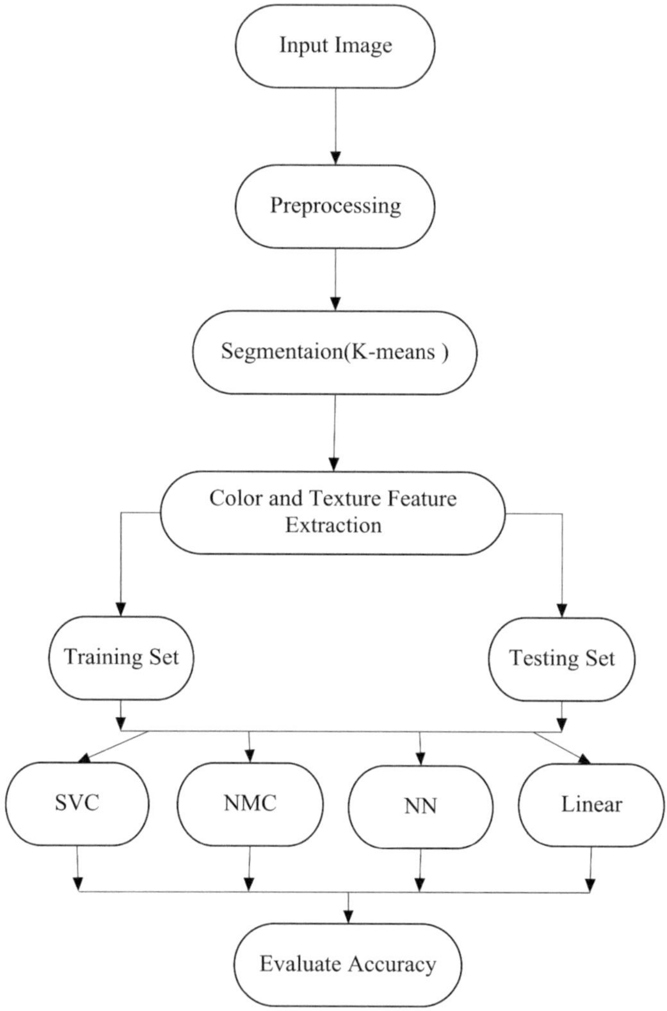

Fig. 1 Proposed method

2.1 Segmentation

Segmentation is a process to partitioning an image into disjoint regions that are homogeneous with respect to a chosen property such as luminance, color, and texture. The aim of segmentation is to change the representation of an image into something that is more meaningful and easier to analyze [30].

K-means Clustering Segmentation: K-means clustering method is one of the simplest unsupervised learning methods and this method is nondeterministic, numerical, and iterative. In this experiment, color images are used as input; hence,

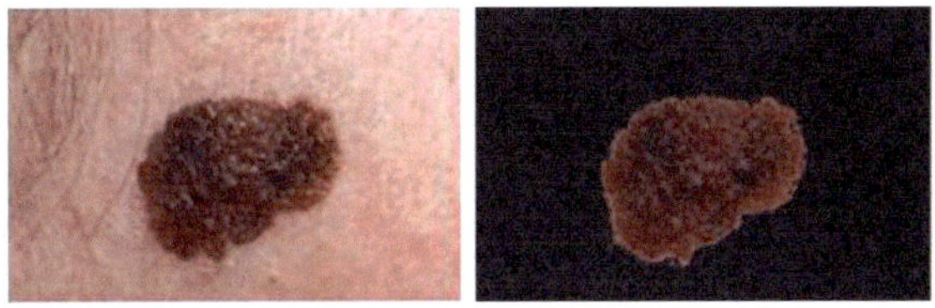

Fig. 2 Input image and segmented image

this technique is used for segmentation. K-means clustering is partitioning method. This method groups objects in the way that within-group variance is minimized. If within-group variance is minimized, then it gives high featured segmented image. The working of this method is as follows [31, 32]:

1. Select K pixels for centers randomly. Initially, center represents group centroids.
2. For grouping the sample data, calculate histogram bin value distance between pixels and selected centroids and assign the group on the basis of nearest distance.
3. Calculate the histogram bin value for new group to find the new position of centroids.
4. If the value of centroids changes, then repeat steps 2 and 3 (Fig. 2).

2.2 *Feature Extraction*

It is important to identify the most effective features to extract from skin cancer lesions and to find the most effective pattern-classification criteria and algorithms for differentiating those lesions.

Local Binary Patterns + Color Percentiles: The LBP feature vector is created in the following way:

1. Divide the window size into cells in the form of n^*n matrix (i.e., 16^*16).
2. If the center pixel value (Threshold Value) is less than the selected pixel, then write 1, otherwise write 0. For rotational invariant features, move in one direction either clockwise or anticlockwise. This step gives 8-digit binary number. Convert it into decimal number for further processing.
3. Compute the histogram of the frequency of each "number" occurring over the given cell.
4. Apply normalization on histograms.
5. Concatenate the histograms of all divided cells.

The combination of LBP + CP presented by Niskanen et al. [33]. In this method, we calculate local binary patterns in an grayscale image. For better results, we reduce the calculated features into rotationally invariant features. Rotational invariance is necessary because when image is rotated, the gray values also rotated in a circular form where origin is same. Then, calculation of feature vector changes with the rotation. We apply discrete Fourier transformation function (DFT) to make features rotationally invariant. There are always 36 rotationally invariant features in a grayscale image. Now, we calculate the color percentiles of first, second, and third quartiles (three points that divide the each channel into four parts) of three channels. Total features from color percentiles are nine. When we concatenate features of these two methods, the final resulting feature vector dimension is 36.

Integrative Co-occurrence Matrices: Integrative co-occurrence matrix [34–36] is a new approach for calculating the inter-channel and intra-channel features. In inter-channel feature calculation, we calculate the co-occurrence matrix features from each of three channels, and in intra-channel, we calculate the features from combination of color channels ((r,g), (r,b), (g,b)). In our experiment, we use five co-occurrence features: energy, contrast, correlation, homogeneity, and inertia. Final resulting feature vector dimension is 30.

Co-occurrence Matrices + Color Percentiles: The combination of co-occurrence matrices and color percentiles is presented by Niskanen et al. [33]. This method is first used for applications in wood inspection. Out of 14 features, we computed five important co-occurrence features for classification. Color percentile is calculated as discussed in LBPnCP method. So, the dimension of final resulting feature vector is 14.

Gabor Features: Gabor features are used to extract scale and orientation information from different channels. We calculate Gabor features for inter-channel of an image with above-mentioned configuration. One channel gives 36 rotationally variant features. So, the resulting feature vector is of 32 + 32 + 32 = 96 dimension when we apply DFT normalization to achieve rotationally invariant features.

Gabor Features + Chromatic Features: Different frequencies and orientations are used in Gabor filters for extracting features from an image. The tuning of Gabor filters with these parameters: number of orientations (n_o) = 6, number of frequencies (n_F) = 4, frequency ratio (F_r) = half-octave, standard deviation of Gaussian envelope (longitudinal) ( ) = 0.5, standard deviation of Gaussian envelop (transversal) ( ) = 0.5, max frequency (F_m) = 0.327. This tuning is same in overall experiment.

Opponent Color LBP: The processing of opponent color LBP feature extraction method is same as LBP method. In opponent color LBP method, we calculate features from inter-channel and intra-channel, but in LBP method, we calculate features from grayscale image. Separate channels give 108 (36 + 36 + 36) features, and paired channels give 108 features. The dimension of resulting feature vector is 216.

Color Ranklets: Ranklet transform is an image processing technique which considers the relative rank of neighboring pixels instead of their values. Invariance against rotation is achieved by computing the discrete Fourier transform (DFT) of a set of ranklet features computed from circular neighborhoods at different rotation angles.

2.3 *Classifier*

Nearest Neighbor and Nearest Mean Classifier: An input pattern is classified in the class of the nearest training pattern (NN) or that of the nearest centroid (NMC) [37, 38]. Absence of parameter tuning makes these classifiers advantageous and easy to implement. In NN, sensitivity to outliers makes it poor performer. In nearest mean classifier, misclassifications arise because probably centroids are not representative.

Linear Classifier: Linear classifiers classify features by making a classification decision which is based on linear combination of the feature values. Linear classifier is originally developed for binary classification. It requires a predefined linear function (hyperplane) that best separates the required classes in the feature space. If the two classes are linearly separable, then perfect separation between classes exists.

Support Vector Classifier: support vector machine is highly effective classifier and currently has a great importance in pattern recognition and artificial intelligence. The tuning of a SVC is very important and needs very careful analysis. In our experiment, we are using RBF kernel function in SVM classifier; then, we have to tune two parameters: C and gamma (the radius of RBF).

C is used during the training phase and says how much outliers are taken into account in calculating support vector. C is a trade-off between training error and the flatness of the solution. If the value of C is larger, then the final training error will be less. But if we increase value of C too much, then the risk of losing the generalization properties of the classifier is high. Large C increases the time needed for training, and small C makes classifier flat. We have to find a C that keeps the training error small and generalizes well. In SVC processing, we choose a kernel function which mapped patterns into a high-dimensional space [39]. According to Hsu et al. [40], if the number of features is not large, then RBF kernel is a reasonable choice.

3 Experiments and Simulation

3.1 *Database Preparation*

In our experiment, we use database of skin cancer images (melanoma and non-melanoma) have been used from University of Waterloo. The University has collected these images from Internet source [41, 42]. Collection of 150 images is used in this experiment. Seventy-five images are of melanoma and 75 of non-melanoma type [43, 44].

Size of image is always an important aspect of image processing experiments. In this experiment, input image is resized to 128^*128, which is quite suitable resolution for obtaining better results.

Resized images are segmentated using k-means clustering, which are trained and tested on different classifiers.

3.2 Result Analysis

The size of feature vector extracted using these feature extraction algorithms is shown below in Table 1. It is clear from the table increasing the dimensionality of feature vector increases the computation time (Fig. 3).

Tables 2, 3, 4, 5, 6, 7, and 8 display classification accuracy of different classifiers for the given feature set using twofold cross-validation approach by varying the training and testing ratio.

Table 1 Elapsed time of methods in increasing order

Method	Number of features	Elapsed time (s)	Reference
LBPnCP	45	54.533748	(M. Niskanen et al. 2001)
GLCM + CP	14	90.964714	(M. Niskanen et al. 2001)
Gabor + CF	36	110.209004	(A. Drimbarean et al. 2001)
Gabor	96	208.859732	(G. Paschos, 2001)
Integrated CM	30	393.141803	(V. Arvis et al. 2004)
Opponent color LBP	216	13781.195905	(T. Maenpaa et al. 2004)
Color ranklets	216	28463.284710	(F. Bianconi et al. 2009)

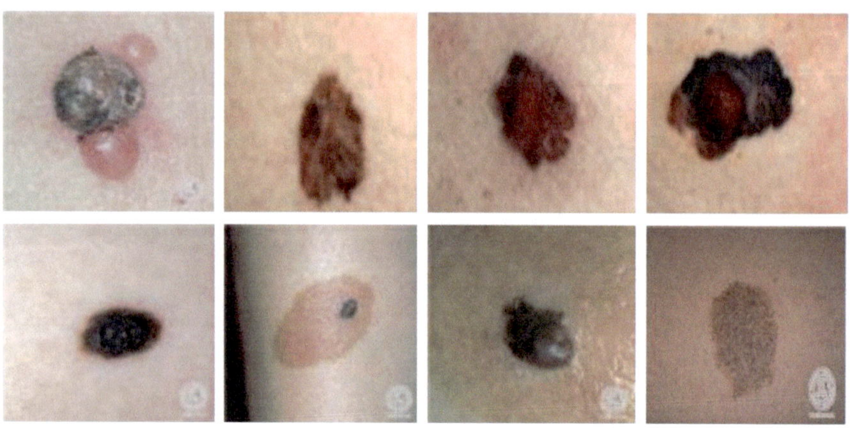

Fig. 3 Images from dataset first row: melanoma, second row: non-melanoma

Table 2 Accuracy of LBPnCP feature extraction technique on different classifiers with given training and testing ratio

Training/testing	NN	NMC	Linear	SVC
(60,15)	65.56	50.00	63.33	62.00
(55,20)	57.14	61.43	51.43	57.14
(50,25)	68.00	58.00	62.00	66.00
(45,30)	80.00	66.67	80.00	83.33
(40,35)	80.00	70.00	93.36	89.14

Table 3 Accuracy of Gabor + CF feature extraction technique on different classifiers with given training and testing ratio

Training/testing	NN	NMC	Linear	SVC
(60,15)	62.00	66.67	63.33	68.89
(55,20)	70.00	77.14	70.00	68.57
(50,25)	76.00	80.00	70.00	82.00
(45,30)	86.67	46.67	86.67	80.00
(40,35)	90.00	80.00	80.00	90.00

Table 4 Accuracy of integrated CM feature extraction technique on different classifiers with given training and testing ratio

Training/testing	NN	NMC	Linear	SVC
(60,15)	70.00	66.67	53.33	66.67
(55,20)	64.29	62.86	67.14	70.00
(50,25)	84.00	80.00	74.00	72.00
(45,30)	86.67	83.33	83.33	86.67
(40,35)	90.00	70.00	80.00	87.76

Table 5 Accuracy of GLCM + CP feature extraction technique on different classifiers with given training and testing ratio

Training/testing	NN	NMC	Linear	SVC
(60,15)	60.00	57.78	58.89	64.44
(55,20)	61.43	47.14	68.57	64.29
(50,25)	74.00	84.00	82.00	82.00
(45,30)	80.00	76.67	83.33	83.33
(40,35)	91.24	60.00	88.65	93.94

Table 6 Accuracy of Gabor feature extraction technique on different classifiers with given training and testing ratio

Training/testing	NN	NMC	Linear	SVC
(60,15)	71.11	74.44	74.44	76.67
(55,20)	64.29	71.43	54.29	67.14
(50,25)	78.00	74.00	78.00	78.00
(45,30)	90.00	83.33	76.67	90.00
(40,35)	90.00	80.00	60.00	96.57

Table 7 Accuracy of opponent color LBP feature extraction technique on different classifiers with given training and testing ratio

Training/testing	NN	NMC	Linear	SVC
(60,15)	55.56	65.56	60.00	63.33
(55,20)	67.14	68.57	64.29	68.57
(50,25)	76.00	66.00	68.00	80.00
(45,30)	86.67	76.67	90.00	86.67
(40,35)	92.29	50.00	80.00	95.11

Table 8 Accuracy of color ranklets feature extraction technique on different classifiers with given training and testing ratio

Training/testing	NN	NMC	Linear	SVC
(60,15)	68.89	57.78	60.00	63.33
(55,20)	72.86	55.71	61.43	70.00
(50,25)	74.00	56.00	72.00	74.00
(45,30)	76.67	60.00	70.00	76.67
(40,35)	80.00	70.00	78.00	80.00

Average accuracies obtained from each classifier for different features are plotted graphically in Fig. 4.

It is clear from Fig. 4 that Gabor feature performs better than the other features. LBPnCP, integrated GLCM, CLCM + CP are the better choice, while the color ranklets and opponent color LBP perform poorly in time and space domains. LBPnCP and Gabor features are best suited for optimum results considering the running time and accuracy of prediction.

Pairwise comparison between the algorithms is done for better visual comparison (Fig. 5). The directions of the arrow denote the method which has more prediction accuracy than the other.

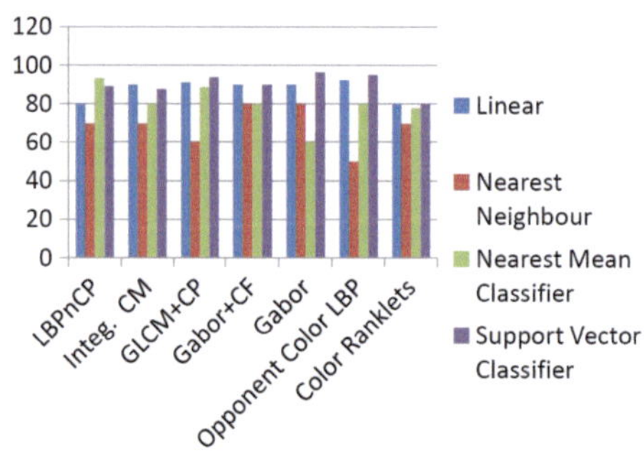

Fig. 4 Average accuracy of different feature extraction algorithms

	LBPnCP	Gabor+CF	Integ. CM	GLCM+CP	Gabor	Opponent Color LBP	Color Ranklets
LBPnCP		↑	↑	↑	↑	↑	←
Gabor+CF	←		↑	↑	↑	↑	←
Integ. CM	←	←		↑	↑	↑	←
GLCM+CP	←	←	←		↑	↑	←
Gabor	←	←	←	←		←	←
Opponent Color LBP	←	←	←	←	↑		←
Color Ranklets	↑	↑	↑	↑	↑	↑	

Fig. 5 Pairwise comparison of the feature extraction algorithm efficiency

4 Conclusion

Experimental results show that color and texture descriptors for skin cancer classification provide good classification accuracy. We have also evaluated the performance of four different classifiers on these extracted features. SVC(support vector classifier) outperforms among all others with the same features set and then comes linear classifier, whose classification accuracy is nearly same as the SVC. 1-NN performs poorly on the given feature sets.

Gabor features proved to be the best features that can be used for this particular application. Other more suitable methods are LBPnCP and integrated GLCM which are also giving the promising results.

Opponent and color ranklets are the methods which are not advisable to be used for the particular application because of their large dimensions and less prediction accuracy as well.

References

1. Linos, E., Swetter, S.M., Cockburn, M.G., Colditz, G.A., Clarke, C.A.: Increasing burden of melanoma in the United States. J. Invest. Dermatol. **129**(7), 1666–1674 (2009)
2. Mayer, J.: Systematic review of the diagnostic accuracy of dermatoscopy in detecting malignant melanoma. Med. J. Aust. **167**(4), 206–210 (1997)
3. Robinson, J.K., Turrisi, R.: Skills training to learn discrimination of ABCDE criteria by those at risk of developing melanoma. Arch. Dermatol. **142**(4), 447–452 (2006)
4. Celebi, M.E., Iyatomi, H., Schaefer, G., Stoecker, W.V.: Lesion border detection in dermoscopy images. Comput. Med. Imaging Graph. **33**(2), 148–153 (2009)
5. Iyatomi, H., Oka, H., Celebi, M.E., Hashimoto, M., Hagiwara, M., Tanaka, M., Ogawa, K.: An improved internet-based melanoma screening system with dermatologist-like tumor area extraction algorithm. Comput. Med. Imaging. Graph. **32**(7), 566–579 (2008)
6. Garnavi, R., Aldeen, M., Celebi, M.E., Varigos, G., Finch, S.: Border detection in dermoscopy images using hybrid thresholding on optimized color channels. Comput. Med. Imaging Graph. **35**(2), 105–115 (2011)
7. H. Zhou, M. Chen, L. Zou, R. Gass, L. Ferris, L. Drogowski, and J. M. Rehg, "Spatially Constrained Segmentation of Dermoscopy Images," Proce. 5th IEEE International Symp. Biomed. Imaging, pp. 800–803, (2008)
8. Schmid, P.: Segmentation of digitized dermatoscopic images by two-dimensional color clustering. IEEE Trans. Med. Imaging **18**(2), 164–171 (1999)
9. Zhou, H., Schaefer, G., Sadka, A., Celebi, M.E.: Anisotropic mean shift based fuzzy C-means segmentation of dermoscopy images. IEEE J. Sel. Top. Sign. Proces. **3**(1), 26–34 (2009)
10. Mete, M., Kockara, S., Aydin, K.: Fast density-based lesion detection in dermoscopy images. Comput. Med. Imaging Graph. **35**(2), 128–136 (2011)
11. R. Melli, C. Grana, and R. Cucchiara, "Comparison of color clustering algorithms for segmentation of dermatological images," Proc. SPIE Med. Imaging Conf. pp. 1211–1219 (2006)
12. Erkol, B., Moss, R.H., Stanley, R.J., Stoecker, W.V., Hvatum, E.: Automatic lesion boundary detection in dermoscopy images using gradient vector flow snakes. Skin Res. Technol. **11**(1), 17–26 (2005)
13. Zhou, H., Schaefer, G., Celebi, M.E., Lin, F., Liu, T.: Gradient vector flow with nean shift for skin lesion segmentation. Comput. Med. Imaging Graph. **35**(2), 121–127 (2011)
14. Abbas, Q., Celebi, M.E., Garcia, I.F.: A novel perceptually-oriented approach for skin tumor segmentation. Int. J. Innovative Comput. Inf. Control **8**(3), 1837–1848 (2012)
15. Celebi, M.E., Aslandogan, Y.A., Stoecker, W.V., Iyatomi, H., Oka, H., Chen, X.: Unsupervised border detection in dermoscopy images. Skin Res. Technol. **13**(4), 454–462 (2007)
16. Celebi, M.E., Kingravi, H.A., Iyatomi, H., Aslandogan, A., Stoecker, W.V., Moss, R.H., Malters, J.M., Grichnik, J.M., Marghoob, A.A., Rabinovitz, H.S., Menzies, S.W.: Border detection in dermoscopy images using statistical region merging. Skin Res. Technol. **14**(3), 347–353 (2008)
17. Wang, H., Moss, R.H., Chen, X., Stanley, R.J., Stoecker, W.V., Celebi, M.E., Malters, J.M., Grichnik, J.M., Marghoob, A.A., Rabinovitz, H.S., Menzies, S.W., Szalapski, T.M.: Modified watershed technique and Post-Processing for segmentation of skin lesions in dermoscopy images. Comput. Med. Imaging Graph. **35**(2), 116–120 (2011)
18. Abbas, Q., Celebi, M.E., Garcia, I.F., Rashid, M.: Lesion border detection in dermoscopy images using dynamic programming. Skin Res. and Technol. **17**(1), 91–100 (2011)
19. Abbas, Q., Celebi, M.E., Garcia, I.F.: Skin tumor area extraction using an improved dynamic programming approach. Skin Res. Technol. **18**(2), 133–142 (2012)

20. Schaefer, G., Rajab, M.I., Celebi, M.E., Iyatomi, H.: Colour and contrast enhancement for improved skin lesion segmentation. Comput. Med. Imaging Graph. **35**(2), 99–104 (2011)
21. Wighton, P., Lee, T.K., Lui, H., McLean, D.I., Atkins, M.S.: Generalizing common tasks in automated skin lesion diagnosis. IEEE Trans. Information Technol. Biomed. **15**(4), 622–629 (2011)
22. V. K. Dehariya, S. K. Shrivastava, R. C. Jain, "Clustering of image data set using K-Means and fuzzy K-Means Algorithms", International conference on CICN, pp. 386–391, (2010)
23. J. F. Aitken, J. Pfitzner, D. Battistutta, O.'Rourke PK, A. C. Green, and N.G. Martin, "Reliability of computer image analysis of pigmented skin lesions of Australian adolescents," J. Cancer, **78**(2), 252–257 (1996)
24. Green, A., Martin, N., Pfitzner, J., O'Rourke, M., Knight, N.: Computer image analysis in the diagnosis of melanoma. J. Am. Acad. Dermatol. **31**(6), 958–964 (1994)
25. H. C. Lee, Skin cancer diagnosis using hierarchical neural networks and fuzzy logic, department of computer science, University of Missouri, Rolla (1994)
26. Seidenari, S., Burroni, M., Dell'Eva, G., Pepe, P., Belletti, B.: Computerized evaluation of pigmented skin lesion images recorded by a videomicroscope: comparison between polarizing mode observation and oil/slide mode observation. Skin Res. Technol. **1**(4), 187–191 (1995)
27. K. Ramlakhan and Y. Shang,: "A mobile automated skin lesion classification system," Proceedings of the 23rd IEEE International Conference on Tools with Artificial Intelligence (ICTAI '11), pp. 138–141, Nov (2011)
28. Cascinelli, N., Ferrario, M., Bufalino, R., et al.: Results obtained by using a computerized image analysis system designed as an aid to diagnosis of cutaneous melanoma. Melanoma Res. **2**(3), 163–170 (1992)
29. F. Bianconia, E. Gonzalezb, A. Fernandezb, S.A. Saet-taa: "Automatic classification of granite tiles through colour and texture features", Preprint submitted to Expert Systems with Applications, Feb 20, (2012)
30. A. Masood and A. Ali Al-Jumaily, "Computer Aided Diagnostic Support System for Skin Cancer: A Review of Techniques and Algorithms" Hindawi Publishing Corporation International Journal of Biomedical Imaging, Volume (2013)
31. V. Jumb, M. Sohani, A. Shrivas: "Color image segmentation using K-Means clustering and Otsu's adaptive thresholding," Int. J. Innovative Technol. Exploring Eng. (IJITEE), February (2014)
32. Ojala, T., Pietikainen, M., Maenpaa, T.: Multiresolution gray-scale and rotation invariant texture classification with Local Binary Patterns. IEEE Trans. Pattern Anal. Mach. Intell. **24**, 971–987 (2002)
33. Niskanen, M., Silven, O., Kauppinen, H.,: "Color and texture based wood inspection with non-supervised clustering", Proceedings of the 12th Scandivanian Conference on Image Analysis. Bergen, pp. 336–342, (2001)
34. Arvis, V., Debain, C., Berducat, M., Benassi, A.: Generalization of the cooccurrence matrix for colour images: application to colour texture classification. Image Anal. Stereology **23**(3), 63–72 (2004)
35. Palm, C.: Colour texture classification by integrative co-occurrence matrices. Pattern Recogn. **37**(5), 965–976 (2004)
36. Bianconi, F., Fernandez, A.: Evaluation of the effects of Gabor filter parameters on texture classification. Pattern Recogn. **40**, 3325–3335 (2007)
37. F. Bianconi, A. Fernandez, E. Gonzalez, and J. Armesto, Robust color texture features based on ranklets and discrete Fourier transform. J. Electron. Imaging **18**, 043012-1-8 (2009)
38. Masotti, M., Campanini, R.: Texture classification using invariant ranklet features. Pattern Recogn. Lett. **29**, 1980–1986 (2008)
39. Scholkopf, B., Smola, A.: "Learning with Kernels: Support Vector Machines, Regularization, Optimization, and Beyond", MIT Press, (2002)
40. Hsu, C., Chang, C., Lin, C.: A practical guide to support vector classification. Department of Computer Science National Taiwan University, Tech. rep. (2010)

41. http://www.dermis.net
42. http://www.dermquest.com
43. B.Gohila vani, A.Selvaraj, M.Vijaya priya, N.Anandh: "Segmentation and Classification of Skin Lesions Based on Texture Features", Int. J. Eng. Res. Appl. Dec. (2014)
44. Maurya R., Singh S.K., Maurya A.K., Kumar A.: "GLCM and Multi Class Support vector machine based automated skin cancer classification IEEE Conference", Page(s):444—447Print ISBN:978-93-80544-10-6 INSPEC Accession Number:14382783, pp. 5–7 Mar. (2014)

Pattern Classification and Retrieval of Content-Based Images—a Step Towards Amelioration

Hemjot and Amitabh Sharma

Abstract The progressive web and computerized advancements have forced endless increment in the measure of visual data accessible to users. This trend prompted the advancement of exploration area where retrieval of images is done through the content of information which became familiar as CBIR (content-based image retrieval). CBIR frameworks are to a great extent utilized as a part of medicinal picture annotation, face recognition systems, security frameworks and so on. In this paper, we will discuss about an efficient system for retrieving images faster since speed and precision are important as well as techniques to obtain better classification of images. To conquer the issue of extensive number of features extracted which obliges vast measure of memory and processing force, we need to build a blend of 3 techniques (SURF, SVM and LDA) which best portray the information with adequate precision. Hence, we are using dimensionality reduction algorithm LDA in combination with SVM for the classification purpose and SURF which is quick and robust interest point detector.

Keywords Content-based image retrieval (CBIR) · Pattern recognition · Classification · Support vector machine (SVM) · Colour histogram · Speeded up robust features (SURF) · Linear discriminant analysis (LDA)

1 Introduction

A significant expansion has been done in the digital field for more than many years. This particular development has motivated research in image databases, which were virtually dismissed simply by conventional systems due to the enormous level of

Hemjot (✉) · A. Sharma
Department of IT, Chandigarh Engineering College, Landran, Mohali
Punjab, India
e-mail: hemso91@gmail.com

A. Sharma
e-mail: cecm.cse.amits@gmail.com

S. Urooj and J. Virmani (eds.), *Sensors and Image Processing*, Advances in Intelligent Systems and Computing 651, https://doi.org/10.1007/978-981-10-6614-6_2

information important to represent images as well as the problems connected with an automatic analysis of images. In order to make use of this kind of huge number of information, efficient as well as effective approaches to get back multimedia details according to their written content ought to be formulated. It was henceforth proposed to build a system which can retrieve images through content information called as content-based image retrieval.

Many techniques were used for building an efficient CBIR system such as SIFT and SVM but the results were not up to the mark. Considering all the drawbacks of previous work done SURF and SVM, implemented in combination could achieve accuracy over 96%. The main focus now became the computation time and the accuracy of image retrieval. An enhancement should be done to achieve the desired results. Therefore, LDA in combination to SURF and SVM was proposed and implemented which came out to be more promising than the existing works. Basic steps included in proposed CBIR system which are necessary for effective and accurate image retrieval are as follows:

1. Image Acquisition is the method of getting digital image information that consists of n range of pictures.
2. The pre-processing level consists of filtering, segmentation and object identification to urge a collection of serious regions and objects.
3. In the stage of Feature extraction, visual data like colour is extracted from the pictures and saves them as feature vectors in a database of feature vectors. One in every of the most important issues with content primarily based image retrieval system is that the sizable amount of options extracted which needs great deal of memory and computation power. To beat this drawback, different algorithms have been proposed. Most promising algorithm for detection, description and matching of the images used is SURF.
   Improved SURF with k-d tree algorithm makes use of integral images in a very efficient way which results into fast detection of interest points and ultimately less time consumption. SURF is used to extract relevant features and descriptors from images, and k-d tree algorithm makes the matching process faster.
4. Next, dimensionality reduction and classification are done. The necessary features are extracted from the database of feature vectors and are stored as reduced feature vector database. Hence, the result of feature extraction phase is a set of features that are reduced to information which is essential to describe an image, and classification is done.
   LDA (Linear Discriminant Analysis) is one of the most popular statistical techniques of pattern recognition and dimensionality reduction. It calculates the linear combination of all the features of the image which separates the class of image. One of the main advantages that LDA provide is to draw a separation line between the different categories and their classification without losing any data.
   Support vector machines (SVMs) are the models which make use of efficient algorithms to analyse data and pattern for the purpose of efficient classification

and analysis. It is a representation of points in space mapped in such a way that the samples belonging to different categories are divided wider by a distinguishing gap further after the dimensionality reduction process.
5. The stage of similarity matching consists of the matching of the reduced feature vectors of input image with the feature vectors of reduced feature vector database and the best matching image is displayed.

2 Literature Review

Anna Wojnar, Antonio M. G. pinheiro [1] depict the medical images retrieval and annotation using a fast descriptor and a classifier. SURF algorithm is used for the extraction of features as it acts as a fast-hessian detector. SVM is used for the feature matching process in combination with the quadratic kernel. **Dong Hui, Han Dian Yuan** [3] explained the more efficient SURF algorithm. The effective image matching method was introduced where the extraction of features is done effectively in the presence of scale rotation in the images. Efficiency is achieved with the introduction of k-d tree algorithm in addition to SURF algorithm. K-d tree helps to search similar points more effectively and quickly. **Herbert Bay, Andreas Ess, Tinne Tuytelaars and Luc Van Gool** [6] demonstrated a combination of detector and descriptor which is invariant towards rotation and scale named as Speeded Up Robust Features. It is best in every aspect as compared to previous schemes and is much faster. **S. Balakrishnama, A. Ganapathiraju** [12] explained the concept of classification technique LDA and its two approaches, class dependent and class independent. Selection of type of LDA is based on the data set and classification problem.

3 Proposed Work

3.1 Loading of Image

In proposed work, three different categories of images are stored in the database that is medical images, face images and random images. Image is loaded after selecting from one category. In Fig. 1(a) different categories are displayed to choose from for the image retrieval process and Fig. 1(b) shows the loading of the selected image.

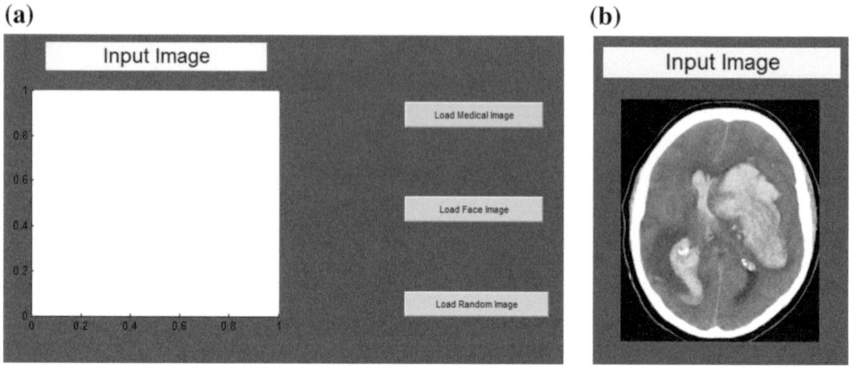

Fig. 1 (**a**) Different categories of images (**b**) loading of image

3.2 *Pre-Processing and Feature Extraction*

Colour Histogram. A histogram is defined as a graph that represents the level of incidence of all the colours present in a picture regardless of the sort of the image. This method portrays the proportion of pixels of every colour in a picture. It's been used as an attribute for feature extraction with the positive aspects like strength with relevance to geometric changes of the objects within the image. More formally, the colour histogram is defined by,

$$\mathrm{h_{A,B,C}(a, b, c) = N.Prob(A = a, B = b, C = c)} \tag{1}$$

Where
A, B and C = R, G, B or H, S, V (three colour channels)
N = number of pixels in the image

The colour histogram is made by separating the colours in an image and keeping a record of pixels of each and every colour. In image retrieval, the colour histogram features from query image are extracted for the process of adding in the feature set matching with all the images in database.

Speeded Up Robust Features (SURF) SURF Descriptor will be used for detection, description and matching of image features.

Detection Interest points will be identified and a vector will be constructed to extract the features. Hessian matrix approximation is used here for the detection of interest point.

$$H(x, \sigma)\begin{pmatrix} L_{xx}(x, \sigma)L_{xy}(x, \sigma) \\ L_{xy}(x, \sigma)L_{yy}(x, \sigma) \end{pmatrix} \tag{2}$$

Where

$$L_{xx}(x,\sigma) = I(x) * \frac{\delta^2}{\delta x^2} g(\sigma) \quad (3)$$

$$L_{xy}(x,\sigma) = I(x) * \frac{\delta^2}{\delta xy} g(\sigma) \quad (4)$$

L_{xx} (x,σ) and L_{xy} (x,σ) are the convolutions of the image with the second derivative of the Gaussian. The use of integral images for calculating convolutions has approximated and speeded up the computations which otherwise is very costly. An integral image I(x) is an image where each point x = (x,y) represents the addition of all the points(pixels) in a rectangular area.

$$I(x) = \sum_{i=0}^{i<x} \sum_{j=0}^{j\leq y} I(x,y) \quad (5)$$

Determinants of hessian matrices are calculated to detect the structures which appear as blobs and are detected at the points where determinants are maximized.

$$Det(H_{approx}) = D_{xx}D_{xy} - (wD_{xy})^2 \quad (6)$$

Where

$$D_{xx} = L_{yy}(x,\sigma) \quad (7)$$

$$D_{xy} = L_{xy}(x,\sigma) \quad (8)$$

Description Haar—Wavelet responses are used to compute the orientation in both directions i.e., x and y. The estimation of the final orientation is done by adding the horizontal wavelet and vertical wavelet responses in the space where an angle of π/3 is within a rotating wedge. The orientation of interest point is then described by choosing the maximum resultant. To assess the primary direction of the feature, Haar Transforms are used.

Matching Process In feature matching, SURF finds out features from same positions in two images and matches two images based on these features. It is a 64-dimensional vector which will adopt nearest neighbour search using a k-d tree.

3.3 Pattern Recognition and Classification of Images

Linear Discriminant Analysis (LDA) LDA (Linear Discriminant Analysis) is the frequently used technique for separation of data and reduction of dimensionality.

Linear Discriminant Analysis smoothly manages the occurrence of sloping and inequality within-class frequencies. This process ensures optimum separation between the classes by maximizing the degree of between-class and the within-class variances in data. The biggest advantage of LDA in proposed work is that it provides large separation of classes without changing the location and also provides a separate boundary between the given classes.

Between-Class Scatter Matrix

$$S_b = \sum_{i=1}^{C} (\mu_i - \mu)(\mu_i - \mu)^T \tag{9}$$

$$\mu = {}^1\!/_C \sum_{i=1}^{C} \mu_i (\text{Mean of entire data set}) \tag{10}$$

Within-Class Scatter Matrix

$$S_w \sum_{i=1}^{C} \sum_{j=1}^{M_i} (y_i - \mu_i)(y_i - \mu_i)^T \tag{11}$$

This transformation is good to retain maximum class separation and thereby reducing the variation caused because of sources other than identity (e.g. illumination).

Support Vector Machine (SVM) Classification and regression of feature points are handled by support vector machine. It is a technique used for binary classification of images but later it was optimized for multi-classification of images and extended for multi-classification.

SVM classifier is trailed by SURF in the whole process. After feature detection, separation is done between matching and non-matching points of the two images. Further, a plane separates the matching and non-matching feature points so as to achieve a clear differentiation between the two images. Therefore, during the matching process, pattern classification is also performed. Arrangement of training examples is done in such a way that SVM training algorithm develops a model that allots new samples into one class or the other according to the training algorithm, making it a non-probabilistic binary linear classifier.

3.4 Similarity Matching of the Images

After the classification of images, next comes out the process of obtaining the best similar image against the query image. The most similar image is displayed with the accuracy rate of similarity which is greater than 98% (Fig. 2).

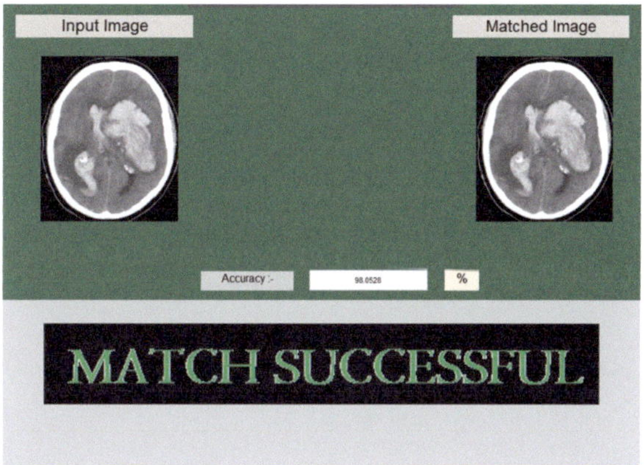

Fig. 2 Matching process

4 Experimental Results

In our research work, main point of achievement is better classification of images with more accuracy and reduced matching time. Therefore, different parameters are compared and evaluated between the previous used algorithm and our proposed combination of techniques to present our achievement.

Different parameters calculated are:

(a) (PSNR) Peak Signal-To-Noise Ratio
(b) (MSE) Mean Square Error
(c) Feature Points
(d) Matching Time
(e) Accuracy (Table 1, Fig. 3).

5 Conclusion

In this paper, a comparative study is presented between two image retrieval methods that are SURF + SVM and SURF + SVM + LDA. One of the real issues with content-based image retrieval framework was the less accurate results and the laborious force taken to retrieve the images. Our proposed combination of algorithm has proved to get an accuracy rate greater than 98%. Improved SURF algorithm with k-d tree has proved to reduce the matching time to a greater extent

Table 1 Comparison between different Parameters based on SURF + SVM and SURF + SVM + LDA

Parameter	Description	Mathematic formula	SURF + SVM (Previous work)	SURF + SVM + LDA (Proposed work)
PSNR	It is the calculation of the ratio of peak signal-to-noise between the input and output images Higher PSNR Value = Better Quality	$PSNR = 20\log_{10}\left(\frac{255}{RMSE}\right)$ Where RMSE is the square root of MSE. Unit of PSNR is in decibels.	3.5267	8.9817
MSE	The MSE (mean square error) is the calculation of cumulative squared error between the two images Lesser MSE Value = Less error	$MSE = \frac{\sum [f(i,j) - F(I,J)]^2}{N^2}$	7.8534e + 03	7.8170e + 03
Feature points	Less feature points with more discriminatory data results in less time consumption and more accuracy	———	544.5000	223.5000
Matching time	The total time is taken to get the most similar image	———	1.7077	1.1887
Accuracy	It measures how accurately the query image matches with the similar image. Achieved accuracy with our proposed method > 98%	———	93.4148	98.4149

and SVM with LDA achieved the purpose of extracting less feature points with a large set of discriminatory information without any loss of data. For better evaluation of results PSNR, MSE, Feature points, Matching time and Accuracy of our proposed work were calculated in comparison with the results of previous work. Thus, the results achieved were promising and best in every aspect.

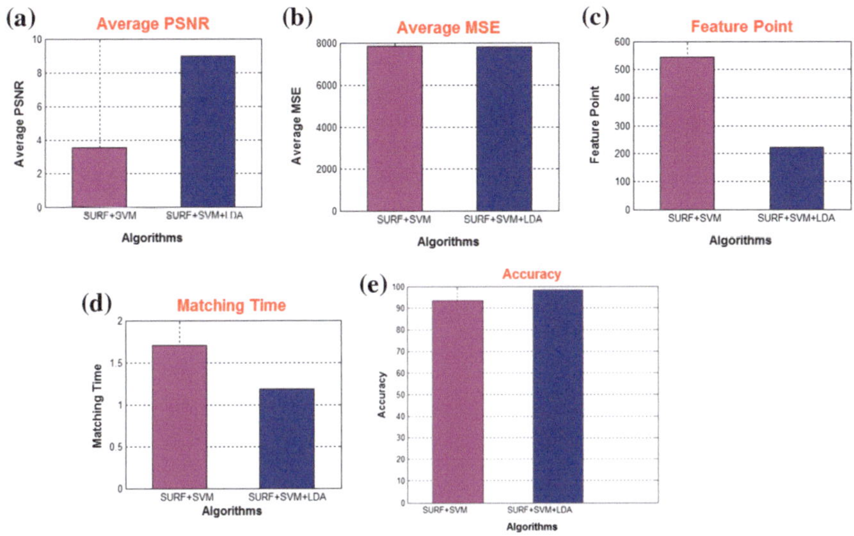

Fig. 3 Comparison between SURF + SVM and SURF + SVM + LDA using different parameters (**a**) PSNR (**b**) MSE (**c**) FEATURE POINTS (**d**) MATCHING TIME (**e**) ACCURACY

References

1. A. Wojnar, Antonio M.G. pinheiro. "Annotation of Medical Images Using The Surf Descriptor" IEEE transactions on medical imaging, **30**(3), 130–134, March (2012)
2. Zhang, B., Vijaya Kumar, B.V.K.: "Detecting Diabetes Mellitus and Non proliferative Diabetic Retinopathy Using Tongue Color, Texture, and Geometry Features". IEEE Trans. Biomed. Eng. **61**(2), 491–501 (2014)
3. D. Hui, H.D. Yuan. "Research of Image Matching Algorithm Based on SURF Features" IEEE Int. Conf. Comput. Sci. Inf. Proc (CSIP) (2012).
4. Dr. Fuhui Long, Dr. Hongjiang Zhang and Prof. David Dagan Feng "Fundamentals of Content-Based Image Retrieval"
5. H.C. Akakin and M.N. Gurcan "Content-Based Microscopic Image Retrieval System for Multi-Image Queries" IEEE tran. inf. technol. biomed. **16**(4), 758–769, July (2012).
6. H. Bay, A. Ess, T. Tuytelaars, and L. Van Gool," Speeded-Up Robust Features (SURF)", Computer Visual Image Understanding, pp 346–359, June (2008).
7. J.R. Uijlings, and A.M. Smeulders, "Real-Time Visual Concept Classification" IEEE Trans. Multimedia, **12**(7), 665–681, November (2010).

8. K.S. Goh, E.Y. Chang, "Using One-Class and Two-Class SVMs for Multiclass Image Annotation"IEEE Trans. Knowl. Data Eng., **17**(10), 1333–1346 October (2005) (Global Journal of Computer Science and Technology, **11**, 10 Version 1.0., May 2011).
9. Lekha Chaisorn, ZixiangFu "A Hybrid Approach for Image/Video Content Representation and Identification" IEEE, (2011).
10. M. Singha, K. Hemachandran, A. Paul, "Content based image retrieval using the combination of the fast wavelet transformation and the colour histogram" IET Image Proc. **6**(9), 1221–1226, (2012).
11. Shaila S G, Vadivel, A.: "Block Encoding of Colour Histogram for Content Based Image Retrieval Applications", pp. 526–533, (2012).
12. S. Balakrishnama, A. Ganapathiraju "Linear Discriminant Analysis—A Brief Tutorial" Institute for Signal and Information Processing Department of Electrical and Computer Engineering, Mississippi State University

Novel Threshold Formulation for Energy Detection Method to Efficient Spectrum Sensing in Cognitive Radio

Rohini S. Kale, Vijay M. Wadhai and Jagdish B. Helonde

Abstract Cognitive radio (CR) is an upcoming technology in wireless communications. Available spectrum used by licensed user is called as primary user (PU). Unused licensed spectrum is called spectrum hole. CR i.e., unlicensed user is called as secondary user (SU). The important task to the CR is to efficiently use the available spectrum without disturbing PU. The success of CR depends on spectrum sensing efficiency. In energy detection method, we have compared the energy of the samples with threshold (denoted as λ). If the sample energy is greater than the threshold, we say the signal is present; otherwise, the signal is absent. This method does not require any prior information of the signal, so energy detection is a popular sensing method due to its simplicity. Through this paper, we would like to propose a novel threshold formulation for energy detection method to efficient spectrum sensing in cognitive radio. Proposed formulation avoids Q inverse function; also, this formulation is independent of number of samples. With rigorous analysis of the proposed formulation for different SNRs, it gives better results compared to other systems.

Keywords Cognitive radio (CR) · Detection probability (Pd) · False alarm probability (Pfa) · Threshold (λ) · Sampled frequency

R.S. Kale (✉)
Department of Electronics and Telecommunication, MIT Polytechnic Pune, Pune, India
e-mail: rohiniskale@gmail.com

V.M. Wadhai
Department of Electronics and Telecommunication, MIT Pune, Pune, India
e-mail: wadhai.vijay@gmail.com

J.B. Helonde
Department of Electrical, ITMCOE Nagpur, Nagpur, India
e-mail: drjbhelonde@itm.edu.in

S. Urooj and J. Virmani (eds.), *Sensors and Image Processing*, Advances in Intelligent Systems and Computing 651, https://doi.org/10.1007/978-981-10-6614-6_3

1 Introduction

Presently, spectrum scarcity is the thirst area in wireless communication. So, the requirement is to utilize intelligently the available spectrum resources [1]. As per the capability of CR, it enables interaction with environment to determine appropriate communication parameters and adapts to the dynamic radio environment. The main task for CR to identify the spectrum usage is called spectrum sensing. Two main methods for spectrum sensing methods are available. They are cooperative spectrum sensing and non-cooperative spectrum sensing. The energy detection method is less complex method. In our system, by novel threshold setting, we have increased the accuracy of our system. Spectrum sensing should be proper. False detection of PU is because of improper prediction of noise. So, noise model matters. Averaging signal and noise PSD's improve Pd and reduce Pfa. In the proposed noise model, shadowing effect, thermal noise, multipath channel fading, and AWGN noise have been considered [1]. PSD's of the signal is compared with the threshold; if it is greater than threshold, then the signal is present; otherwise, signal is absent. Detection probability of our system is increased. This paper is organized as follows: Sect. 2 explains the related work. Section 3 gives system model. Section 4 describes simulation results. And finally, the conclusion is drawn in Sect. 5.

2 Related Work

Daniel Mercedes Martinez Plata [2] proposed a system based on the dynamic detection of detection threshold. They have used energy detection method for spectrum sensing. For threshold selection, they have considered constant false alarm rate (CFAR) method. In this method, selection of threshold is carried out with present conditions of noise levels. For detection of signal, they have considered QPSK modulated signals. For simulation, they have set pfa = 0.2. Simulation result shows the number of samples increases from 100 to 1000, detected signal power higher than −63 dBm. They have detected the signal having SNR in the range from −4 dB to −20 dB.

Reference [3] FCC gives IEEE 802.22, the first standard based on the cognitive radio technology. Reference [4, 5] gives fixed spectrum assignment policy and geographical variations; the available spectrum used only from 15 to 85%. Reference [6] gives definition of white spaces. The unused frequency spectrum in licensed band is called white spaces. Reference [7] has implemented issues of the MRSS techniques. The effects of inphase (I) and quadrature (Q) gain and phase mismatch with various modulation types. Reference [8] has implemented prototype signal detector in GNU radio environment. Their signal detection algorithm is based on eigenvalue. Reference [9] has performed their analysis with two Sus for energy consumption and transmission delay for green communication context. They have performed their analysis with distributed collaborative sensing. Reference [10] has performed cognitive radio-enabled hospital system. They have done their analysis with dynamic spectrum access and sensor devices. With the consideration of above-related work, it is necessary to sense spectrum properly to avoid false detection. So, we propose an algorithm for spectrum sensing in cognitive radio using novel threshold formulation. Proposed algorithm will minimize the false detection probability and will increase the spectrum sensing efficiency. Hypothetically, energy detection can be considered as a method based on binary decision [11]. This can be written as follows

$$Y_s(t) = n(t) \quad H_0 \tag{1}$$

$$Y_s(t) = s(t) + n(t) \quad H_1 \tag{2}$$

where *s* (*t*) is the received signal and *n* (*t*) is the Additive White Gaussian Noise (AWGN), shadowing noise and thermal noise. H_0 and H_1 represent the two outcomes of the energy detection method. The decision threshold λ is an important aspect since it represents the parameter configured by the system designer to control the spectrum sensing performance. The knowledge of noise power and signal power is necessary to get optimum value of threshold λ. Noise power can be estimated to get the signal power [11, 12].

$$\lambda = Q^{-1}\left(Pfa\sqrt{2N} + N\right)\sigma_n^2 \tag{3}$$

where Q function is the tail probability of the standard normal distribution. N is number of samples. Pfa is false alarm probability, Pd is detection probability, and σ_n is noise variance.

To evaluate the performance of sensing method, the following terminologies are adopted:

1. Probability false of alarm detector

$$P_f = P_r\{E > \lambda\}| \quad H_0 \tag{4}$$

2. Probability of detection

$$P_d = P_r\{E > \lambda\}| \quad H_1 \tag{5}$$

3 System Model

Parameters selected for system model:

Number of users (PU) = 1–10.

Type of radio = TVUHF.

TVUHF spectrum frequency range = 470–890 MHz TVUHF. Spectrum bandwidth = 6 MHz.

Total channels = 70.

Modulation type = 16 QAM.

Noise considered: AWGN, Shadowing noise, and thermal noise.

Traffic model:

Traffic generation by Poisson distribution. A call generated is of 120 calls/hour.

$$y = \frac{\lambda^x}{x!} e^{-\lambda} \tag{6}$$

Exponential distribution for Holding time. It is 180 ms.

$$y = \frac{1}{\mu} e^{\frac{-x}{\mu}} \tag{7}$$

Noval Threshold Formulation (R_K Threshold Formulation)

The formulation is independent of Q inverse function. Complexity of this formulation gets reduced. This formulation is independent of samples. We get threshold (λ_m) as follows.

$$\lambda_m = 2\pi m_f \sigma \tag{8}$$

$$T_1 = 10 \log_{10} \lambda_m \tag{9}$$

$$m_f = 0.005 \times SNR + 0.48 \tag{10}$$

σ = normfit (averaging over N PSD)

T_1 = threshold power in dB

$$y = f(x \mid \mu, \sigma) = \left(\frac{1}{\sigma\sqrt{2\pi}}\right) e^{\frac{-(x-\mu)^2}{2\sigma^2}} \tag{11}$$

$$\sigma^2 = \frac{1}{n} \sum_{i=1}^{n} (x_i - \bar{x})^2 \tag{12}$$

$$\bar{x} = \sum_{i=1}^{n} \frac{x_i}{n} \tag{13}$$

Algorithm

Inputs: Spectrum (TV-UHF), Sampling Frequency (fs), Sampling time (Tw), SNR ***Output:*** Threshold (λ)

Procedure:
Declaration:

$$TVUHF = \{F \,|\, 470e6 \leq F \leq 890\,e6\} \tag{14}$$

$$f = \left\{f_k \,|\, (f_{k+1} - f_k) = 6e^6,\ f_k \in TVUHF\right\} \tag{15}$$

$$t\_Int\{t_i \,|\, (t_{i+1} - t_i)\} = \frac{1}{f_s} \tag{16}$$

t_Int = time interval

1. Get Shadowing effect Noise (N_{se})
2. Get Multipath Channel fading Noise (N_{rfc})
3. Get White Gaussian Noise $\left(N_{awgn}\right)$
4. Get Thermal Noise $(N_{thermal})$
5. $Sig\_Noise = N_{se} + N_{rfc} + N_{awgn} + N_{themal}$
6. $\forall f_k(f_k \in F \rightarrow f_k \in TV\_UHF)$ calculate
   P_{fk} using *Discrete* time integrator
   P_{fk} is power at frequency f_k
   Trapezoidal Rule Integration (Discrete time integral)

$$f_k \in F$$

$$t \in t\_\mathrm{int}$$

$$\mathrm{PSD_{noise}} = \left[\frac{\frac{1}{T_w}\int_{\infty}^{t\_\mathrm{int}} (\text{noise wI})}{N}\right] \text{averaging over NPSD} \tag{17}$$

$$\mathrm{PSD_{signal}} = \left[\frac{\frac{1}{T_w}\int_{\infty}^{t\_\mathrm{int}} (\text{signal wI}))}{N}\right] \text{averaging over NPSD} \tag{18}$$

7. Using Normal Distribution calculate mean (μ) and standard deviation (σ) for $\mathrm{PSD_{noise}}$

$$[m\mu,\ \text{Sigma}] = \text{normfit}\ (\mathrm{PSD_{noise}})$$

8. Calculate Threshold (λ_m) using following equation

$$\lambda_m = 2\pi m_f \sigma. \tag{19}$$

9. Calculate T

$$T = 10\log_{10}(\text{PSD\_signal})$$

T = Received signal power in dB

if
$T > T_1$ primary user is present
else
primary user is absent

4 Simulation Results

For simulation, we have used:

1) Used MATLAB R2012a
2) Run on machine with configuration:
Intel(R) Core (TM) i5-2410 M
CPU 2.30 GHz processor
64-bit operating system
4 GB RAM

Graphical user interface of the system is as shown in Fig. 1. As per R_K threshold formulation, sensed TVUHF spectrum for different users at different SNRs from −20 dB to +20 dB was analyzed. Figure 2 gives detection of signal at number of users is 9. As per our Novel R_k threshold formulation, in Eq. (10),

Proposed formulation have also analyzed for different number of users at different SNRs from −20 dB to +20 dB. Figure 4 gives detection of different users at different SNRs. As per the analysis, till five users, we have detection of 100%, and at higher SNR with more users, we have detection from 98 to 90%.This analysis is shown in Fig. 3.

$$C1 = 0.005 \times SNR + 0.45 \tag{20}$$

$$C2 = 0.005 \times SNR + 0.40 \tag{21}$$

$$C3 = 0.005 \times SNR + 0.35 \tag{22}$$

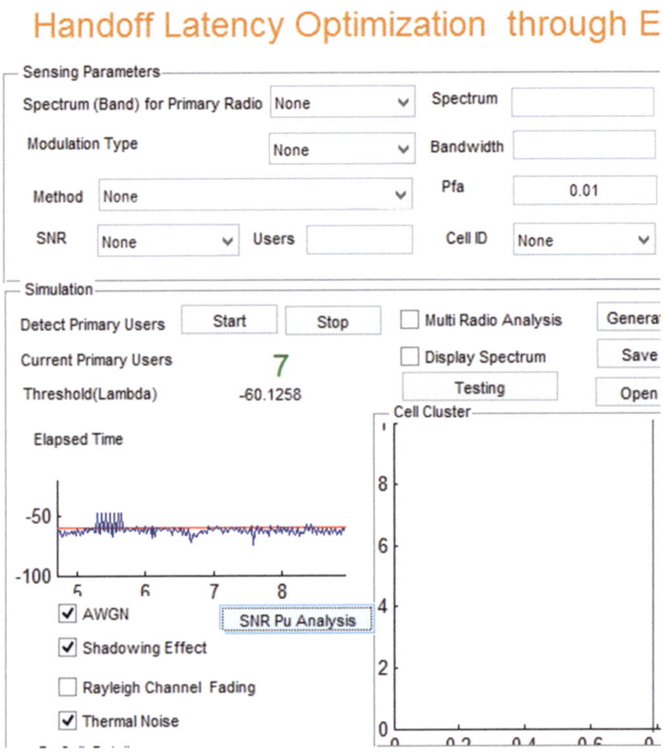

Fig. 1 GUI for Spectrum Sensing

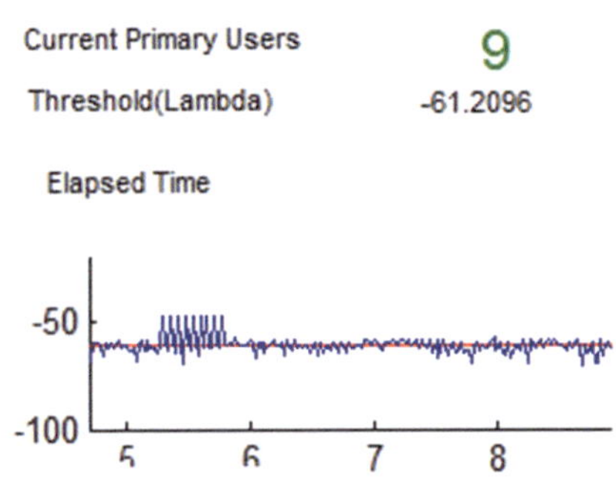

Fig. 2 Sensed TVUHF spectrum at user = 9 and SNR = −20 dB to 20 dB

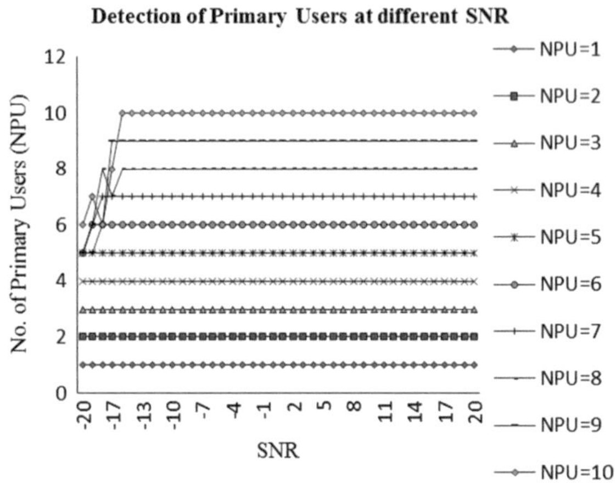

Fig. 3 Detection of different Users at different SNRs

$$C4 = 0.005 \times SNR + 0.48 \tag{23}$$

$$C5 = 0.005 \times SNR + 0.50 \tag{24}$$

$$C6 = 0.005 \times SNR + 0.55 \tag{25}$$

As per Table 1, percent detection analysis has been done, as shown in Fig. 4. C4 gives best results; its statistical equation is displayed on the graph, whose regression value (R^2) is 0.96.So, we have selected constant 0.48 in the mf equation. In the proposed R_K threshold formulation, why we preferred coefficient for mf is 0.005? We have considered different coefficients in equation no. (26) to (29). We have analyzed these equations for percent detection of different number of primary users.

Table 2 gives percent detection of number of users. Analysis in Table 2 indicates Y1 i.e., Equation (26) gives best percent detection.

Table 1 Percent detection at different constant

No. of Users	Percent detection for different constant					
	C1	C2	C3	C4	C5	C6
1	100	100	88	100	100	100
2	100	100	95	100	100	100
3	100	95	93	100	100	98
4	100	93	83	100	100	95
5	98	98	73	100	98	95
6	95	90	65	98	98	93
7	83	85	60	96	75	85
8	90	88	73	93	85	90
9	93	88	60	93	88	85
10	88	88	60	90	88	85

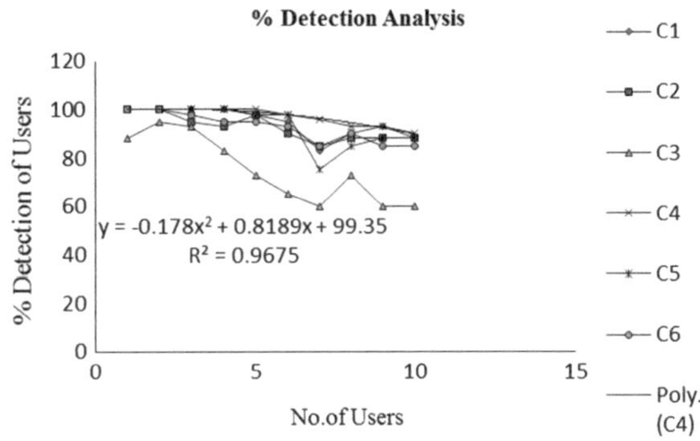

Fig. 4 Percent detection analysis

Table 2 Percent detection analysis for diff. coefficients

No. of Users	Percent detection accuracy for different coefficients			
	Y1	Y2	Y3	Y4
1	100	100	100	100
2	100	100	100	100
3	100	100	100	100
4	100	98	95	98
5	100	95	98	98
6	98	95	95	100
7	96	93	93	98
8	93	93	80	98
9	93	88	90	90
10	90	88	88	93

$$Y1 = 0.005 \times SNR + 0.48 \tag{26}$$

$$Y2 = 0.004 \times SNR + 0.48 \tag{27}$$

$$Y3 = 0.003 \times SNR + 0.48 \tag{28}$$

$$Y4 = 0.006 \times SNR + 0.48 \tag{29}$$

As per Table 2, percent detection analysis has been done, as shown in Fig. 5. Y1 gives best results; its statistical equation is displayed on the graph, whose regression value (R^2) is 0.967. So, we have selected coefficient 0.005 in the mf equation.

Table 3 gives comparison of proposed Novel R_K threshold formulation with previous work.

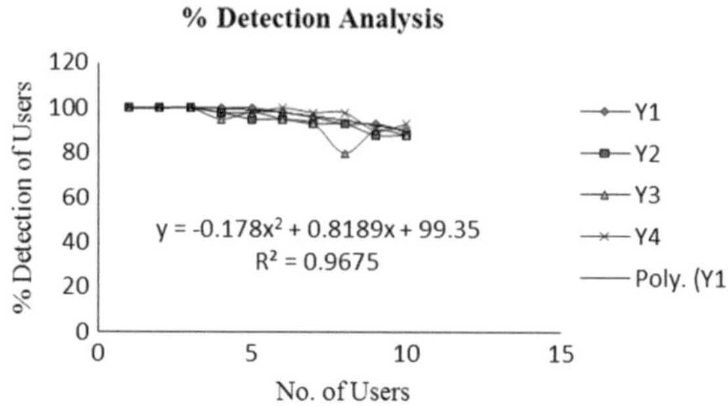

Fig. 5 Percent detection analysis at different coefficients

Table 3 Comparison with previous work

Previous work [10]	Previous threshold formulation [4]	Proposed novel R_K threshold formulation
$\lambda = ((Q^{-1}(pfa)\sqrt{2N}) + N))\sigma_n^2$	$\lambda_n = ((Q^1(pfa)(\sqrt{2N} + N))\left(\sqrt{\frac{\sigma}{100}}\right)^2 x\zeta^2$	$\lambda_m = 2\pi m_f \sigma$
Pfa = 0.2	Pfa = 0.01	Independent of pfa
N = 100–1000	N = 10–4000	Independent of samples
SNR = −4 dB to −20 dB	SNR = 0 dB to −15 dB	SNR= −20 dB to 20dB (up to 6 users)

5 Conclusion

In the proposed R_K threshold formulation rigor analysis at different SNRs, we have 100% detection till five users. As the number of users is increased from six users to ten users, again we have detection of 100% at SNRs from −16 to +20 dBs. At higher noise, i.e., for −20 to −17 dB SNR, users are more than six; then, we have detection from 90 to 97%. Average accuracy of spectrum sensing in TVUHF is 97%. As the Q inverse function was avoided, it is independent of number of samples. So, complexity of our system reduces.

Future Scope: Future plan will be to design and implement the simulated work for spectrum sensing using hardware. Also, performance analysis of this designed system will be a part of the future plan.

References

1. Kale, R.S.: "Efficient spectrum sensing in cognitive radio using Energy Detection method using time domain", ijrat vol.2, No.4, April 2014 E-ISSN:2321-9637

2. Daniela Mercedes Martinez Plata: "Evaluation of energy detection for spectrum sensing based on the dynamic selection of detection threshold". SciVerse Science Direct, 135–143 (2012)
3. The Federal Communications and Commissions, FCC 04-113 IEEE 802.22 Working Group on Wireless Regional Area Networks, IEEE P802.22™/D0.1, May 2006
4. FCC Spectrum Policy Task Force, November 2002, *ET Docket No. 02135*, Washington DC
5. Cabric, D., Mishra, S., Willkomm, D., Broderson, R., Wolisz, A.: "A cognitive Radio approach for usage virtua unlicensed spectrum", White paper 2004
6. Shukla, A.: "Cognitive Radio Technology: A Study for of com – Volume 1." QINETIQ/06/00420 Issue 1.1, February 12th, 2007
7. J. Park: "Implementation Issues of Wide Band Multi Resolution Spectrum Sensing (MRSS) Technique for Cognitive Radio (CR) System", IEEE 2006
8. Sarijari, M., Marwanto, A.: "Energy detection sensing based on GNU radio and USRP, An analysis study", 2009 IEEE 9th Malaysia International Conference on Communications (MICC), No.15–17, pp. 338–342, Dec. 2009
9. Elmachkour, M.; ENSIAS, "The greening of spectrum sensing: a minority game-based mechanism design" Communications Magazine, IEEE, Vol. 52, Issue 12, 11 December 2014
10. Al Mamoon, I.: "Definition, design and priority management of a cognitive radio driven hospital: CogMed" Biomedical Engineering and Sciences (IECBES), 2014 IEEE Conference on, 8–10 Dec. 2014
11. Kale, R.S., Helonde, J.B., Wadhai, V.M.: "Efficient Spectrum Sensing in Cognitive Radio Using Energy Detection Method with New Threshold Formulation", Kerala, IEEE conference icmicr 2013, 4–6 June 2013
12. Nisha, Y.: "A Comprehensive Study of Spectrum Sensing Techniques in Cognitive Radio", International Journal of Advances in Engineering & Technology, July 2011. ©IJAET ISSN: 2231-1963

A Novel Architecture to Crawl Images Using OAI-PMH

Shruti Sharma, Parul Gupta and C.K. Nagpal

Abstract With this evolving size of WWW, it is not possible for search engines to crawl and index the whole Web. Therefore, intelligent crawlers are needed that crawl only those sections of the WWW which contains preferred information. Image crawling is also such a technique that requires to crawl and index special type of information related to the images from the Web. Unlike traditional image crawlers, in this chapter, a novel method for crawling the images from web pages as well as from pdf documents has been proposed that not only uses metatags related to the image but also uses some important features that are related to the images. This proposed work also reports very promising outcome.

Keywords Search engine · Image crawler · Web crawler · Hidden web · OAI-PMH

1 Introduction

Search engine's users expect the results of the search to be relevant and efficient. With this very fact, competence in this domain is growing rapidly. Most of the search engines are text based. However, Web is an amalgamation of text, images, multimedia object, etc. Search engine generally uses a collection of information to answer all user queries. It uses Web crawlers that download web pages and add them to the search engine's collection. The collection thus created is indexed for proficient execution of user queries.

Shruti Sharma (✉) · Parul Gupta · C.K. Nagpal
YMCA University of Science and Technology, Faridabad, India
e-mail: shruti.mattu@yahoo.co.in

Parul Gupta
e-mail: parulgupta_gem@yahoo.com

C.K. Nagpal
e-mail: nagpalckumar@rediffmail.com

S. Urooj and J. Virmani (eds.), *Sensors and Image Processing*, Advances in Intelligent Systems and Computing 651, https://doi.org/10.1007/978-981-10-6614-6_4

An image crawler also works in the similar fashion. It crawls the Web, extracts images and adds them to the image search engine's repository [1, 2 and 3]. On the other hand, images do not exist only as the part of the Web documents. They are also present contained by pdf documents. WWW contains plenty of research documents in the form of pdf documents. These contents are not being considered by the existing crawlers. Although many search engines index pdf documents, but they fail to do so if the properties (like title, author and keywords) of the document are not present in the document itself.

If a user searches for an image, then its related keywords need to be specified. An image search engine uses matching algorithm to find images related to that keyword. However, text cannot be matched with an image. The proposed work targets to extract images from web pages along with the metadata about these images. This metadata is extracted from <img> tag, surrounding text, caption of the image, title of the page, metatag of the page, etc.

The work done mainly focuses on the following two issues:

How to extract images along with the metadata from web pages.

How to extract images along with the metadata within the pdf documents [4].

Out of these images and their metadata, a metadata repository is created, which is updated at regular intervals. This metadata repository content is disseminated using OAI-PMH which has been discussed later in this article.

This chapter consists of the following sections: Sect. 2 contains the work in the area of image crawling; Sect. 3 provides the proposed work of image crawling; Sect. 4 shows the results and analysis related to proposed work; and Sect. 5 concludes the paper.

2 Related Work

Many research groups have worked in the area of image retrieval [1, 5, 6]. Most of which is focused on the images retrieved from web pages. Furthermore, PDF documents are impenetrable part of traditional Web crawlers. Jasco (2001) et al. state that Adobe PDF files represent a large domain of the Web that is invisible to traditional search engines, and thus, to users [4]. Indexing of pdf documents is complex, and this issue cannot be ignored.

AMORE [6] is Advanced Multimedia Object Retrieval Engine. It has a crawler module that extracts images from the Web. The results are searched by matching the query words with metadata related to the images. Thus, the keyword-based mining becomes equally important. The proposed work targets keyword extraction and thus makes search results more relevant. The concept of text descriptors and visual descriptors for images may also be used [7].

Atlas WISE [5] is a Web-based image retrieval engine. This service functions on image gathering, conversion to ppm format, thumbnail creation, feature extraction, histogram computation, etc. This article falls short in answering the following:

1. What does image data contain?
2. How the data about the image are collected?

A review for the available literature shows the following short comings in the current research work:

1. In precise, pdf documents are less indexable [7–9] and so are the images inside them. A more tangible technique needs to be devised to automate the process of metadata extraction of images present in web pages as well as PDF documents.
2. Web is vast, and to get high-quality results, the coverage of the content available over it needs to be maximized. In fact, there has to be some mechanism so that more of surface Web and pdf documents can be processed.
3. Images over Web exist in various formats, e.g. jpg, bmp and ico. Also images exist in different dimensions.
4. Icon images, water marks, etc. also exist over WWW, however they are ineffectual images; hence, we need to ignore them.

Therefore, in order to avoid the above-mentioned drawbacks, an image crawler needs to be designed that can download the variety of images from Web and pdf documents and also effectively relate the metatags and other relevant keywords to the corresponding images.

3 Proposed Architecture of an Image Crawler

In this work, an approach to jointly target web pages as well as pdf documents for image collection has been proposed. First of all a metadata repository is constructed that comprised of the images and their extracted metadata content as shown in Fig. 1. Over this metadata repository, with the help of OAI-PMH, HTTP request is issued to automatically convert the metadata into XML pages. These newly

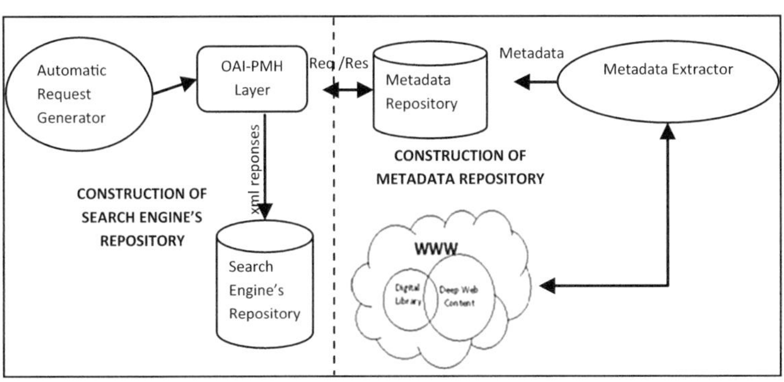

Fig. 1 Functional block diagram of AIC

generated XML pages are then added to the search engine's repository where they are indexed. The entire process is divided into two phases as depicted in Fig. 1:

Phase 1: construction of metadata repository
Phase 2: construction of search engine's repository

The construction of metadata repository phase involves various other modules. The metadata repository is built up of the images and the data about these images. These images can be mined either from the web pages or from the PDF documents. The surrounding text, page title, keywords, image caption, ALT tag, etc. are taken into consideration for generating the metadata about an image. Apart from image and its describing data, a track of some more image-related belongings like size of the image and URL of the page at which image appeared.

The construction of search engine's repository phase involves an automatic request generator. The task of this module is to issue an HTTP request using OAI-PMH [10] over the metadata repository constructed in phase I. In response to the HTTP request, various XML response pages are generated, which are added to the search engine's repository.

The OAI is an initiative to develop and promote interoperability standards that aim to facilitate the efficient dissemination of content. It aims to harvest selectively from data repositories. It provides an application-independent interoperability framework based on metadata harvesting. There are two classes of participants in the OAI-PMH framework as shown in Fig. 2.

Data providers administer systems that support the OAI-PMH as a means of exposing metadata, and service providers use metadata harvested via the OAI-PMH as a basis for building value-added service.

OAI-PMH supports six verbs [11] shown in Table 1 that can be used to query the metadata repository. The request component creates HTTP requests using these verbs and sends them to the data providers. These data providers thus generate the well-formed XML response pages in reply to these requests that comprise of repository's metadata records.

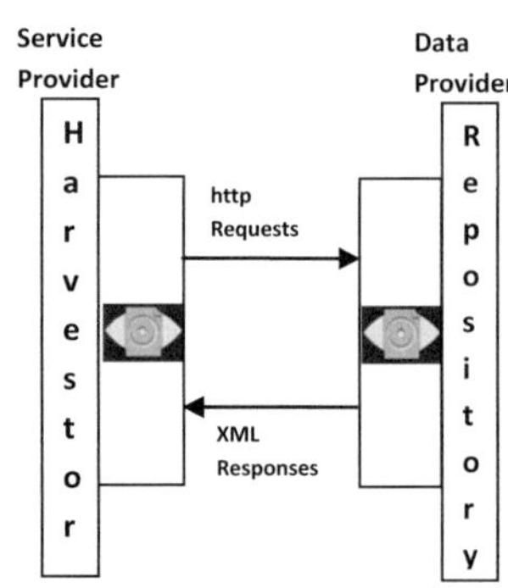

Fig. 2 Block diagram of OAI-PMH

Table 1 OAI-PMH verbs

Verb	http Request Format
GetRecord	http://arXiv.org/oai2?verb=GetRecord&identifier=oai:arXiv.org:cs/0112017&metadataPrefix=oai_dc
Identify	http://memory.loc.gov/cgi-bin/oai?verb=Identify
ListIdentifiers	http://an.oa.org/OAI-script?verb=ListIdentifiers&from=1998-01-15&metadataPrefix=oldArXiv&set=physics :help
ListMetadataFormats	http://www.perseus.tufts.edu/cgi-bin/pdataprov?verb=ListMetadataFormats&identifier=oai:perseus.tufts.edu: Perseus:text:1999.02.0119
ListRecords	http://an.oa.org/OAI-script?verb=ListRecords&from=1998-01-15&set=physics:hep&metadataPrefix=oai_rfc 1807
ListSets	http://an.oa.org/OAI-script?verb=ListSets

3.1 Phase I: Construction of Metadata Repository

It is the first phase of the proposed work that is used for extracting the metadata related to the images and creating the metadata repository that is later used in phase II. It consists of the following components as represented in Fig. 3:

1. Dispatcher(): This component reads the URLs from the database and fills the SeedURLQ. It may also get initiated by the user who provides a seed URL in the beginning. It sends a signal called *SomethingToCrawl* to the Classifier. Its algorithm is given below:

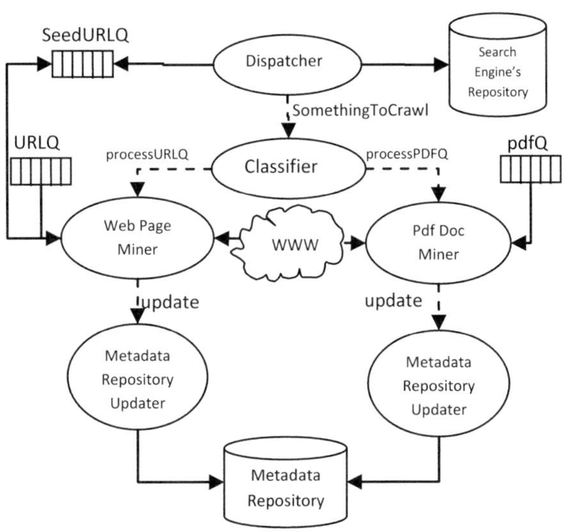

Fig. 3 Phase I—construction of metadata repository

```
Dispatcher()
Begin
Do forever
While(SeedURLQ is not FULL)
Read URLs from Search Engine's Repository;
Add them to SeedURLQ;
Signal(SomethingToCrawl);
}
End
```

2. Classifier(): The task of this component is to categorize a URL into pdf link or pagelink based on its content. This module waits for a signal, *SomethingToCrawl* from the dispatcher to start its work. After classifying the URLs, it is the task of this process to add them to URLQ and pdfQ and to signal the corresponding downloaders as well. If it is a web page, then a signal, *ProcessURLQ*, is sent to the WebPageMiner() module. If it turns out to be a pdf document, then a signal, *ProcesspdfQ*, is sent to the pdfDocMiner module. The algorithm for a Classifier is as follows:

```
Classsifier()
Begin
Do Forever
{
Wait(SomethingToCrawl);
While(SeedURLQ is not Empty)
{
Pick URL from SeedURLQ;
Assign it to URLQ or pdfQ accordingly;
Signal (ProcessURLQ) or
Signal(ProcesspdfQ) accordingly;
}
}
End
```

3. PdfDocMiner(): The task of this component to download all the pdf documents. It waits for a signal, *processpdfQ*, from the Classifier. It processes the pdfQ. It removes one link from the pdfQ and downloads the corresponding document. The content of the pdf file is then mined for images and their related data. The data collected thereof is updated in the metadata repository by signalling *update* to the UpdateRepository() module. Its algorithm is given below:

```
pdfDocMiner()
Begin
```

```
Do Forever
Wait(processpdfQ);
While(PDFQ is not empty)
{
Download the pdf document corresponding to the
URL;
Extract all images in the pdf file;
Extract the metadata about the images from the
keywords and figure Captions;
}
Signal (update);
}
End
```

4. WebPageMiner(): The task of this component is to download all URLs one after the other in the URLQ. This module functions in the similar way as pdfDocMiner(). The two major differentiating points are first, pdfDocMiner() works with pdf documents, on the other hand, WebPageMiner() works with general web pages. Secondly, WebPageMiner() also extracts the hyperlinks present on a page and adds them to the SeedURLQ.

```
WebPageMiner()
Begin
Do Forever
{
Wait(processURLQ);
While(URLQ is not empty)
{
Download the contents of the URL;
Extract all images on the page;
Extract the metadata about the images;
Extract all the links and add them to the
SeedURLQ;
}
Signal (update);
}
End
```

5. UpdateRepository(): This process waits for the signal update from pdfDocMiner () and webPageMiner() modules and on receiving the same; it updates the metadata repository with the content of the buffer. The algorithm of Update Metadata Repository component is given below:

```
UpdateRepository()
Begin
Set MaxSize to the maximum size of a batch;
Do forever
Begin
Wait (update);
No-of-records = 0;
While (DocBuffer is not empty & No-of-records <
MaxSize)
{
Pickup an element from DocBuffer;
Compress the document;
Add to the batch of records to be updated;
No-of-records = No-of-records+1;
}
Update batch to database;
End
End.
```

3.2 *Phase II: Construction of Search Engine's Repository*

In this phase, the metadata repository created by the metadata extractor is used to construct the database (XML files) for the search engine which is later indexed for relevant information retrieval. This phase has the **AutomaticRequestGenerator()** module to automatically issue timed and regular OAI-PMH requests based on the six verbs to the OAI-compliant repository and add the resultant XML response pages of the metadata to the search engine's database. The HTTP requests are generated on the basis of selective harvesting approach based on datestamps and sets. The six OAI-PMH requests with format are shown in Table I.

The Apache's tomcat has a built-in module mod_oai [12] to provide support for these six harvesting verbs of OAI-PMH.

4 Results & Experimentation

The proposed work was implemented in Java to crawl various Web and pdf documents over www. Figure 4 shows the snapshot of the implementation. For example, here http://www.yahoo.com/index.html is given as a seed URL to be processed. All the images on this page with their respective mined keywords were added to the metadata repository. The same process was repeated for various other websites and pdf documents.

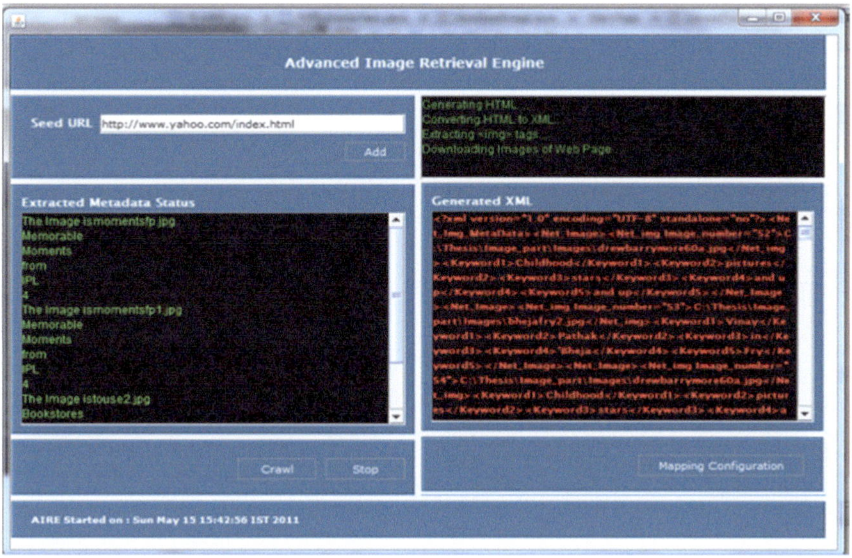

Fig. 4 Snapshot of proposed work (AIC)

During the first crawl, a metadata repository with 1108 images from web pages and 156 images from pdf documents was created [13, 14]. As many as 3424 token for web pages and 637 tokens for pdf documents were generated. The preprocessing of the metadata, after removing the duplicate tokens and stop words, generated, 2167 keywords from the web pages and 458 keywords from the pdf documents. The manual analysis of the images and their extracted metadata revealed that for web pages 1724 keywords and for pdf documents 387 keywords were correctly related with their corresponding images. The accuracy of the entire system was calculated as:

$$Accuracy = \frac{\text{Relevant Keywords}}{\text{Total Extracted words}}$$

Therefore, Accuracy of the first crawl comes to be:

For Web Docs:

$$\text{AW}_1 = \frac{1149}{1512} * 100 = 75.99\%$$

For pdf Docs

$$\text{AP}_1 = \frac{234}{293} * 100 = 79.86\%$$

In order to remove biasing in analysis of proposed work, a second round of crawl with different set of seed URLs was performed. From the Table 2, the accuracy was again computed for the second crawl as:

Table 2 First, Second and third crawl Findings

	First crawl		Second crawl		Third crawl	
	Web pages	PDF docs	Web pages	PDF docs	Web pages	PDF docs
Number of images	732	82	937	96	1108	156
Number of tokens	2143	346	2745	431	3424	637
Number of keywords	1512	293	1917	367	2167	458
No of relevant keywords	1149	234	1552	301	1724	387

$$\text{For Web Docs}: \quad AW_2 = \frac{1552}{1917} * 100 = 80.95\% \qquad \text{For pdf Docs} \quad AP_2 = \frac{301}{367} * 100 = 82.01\%$$

Consequently, the accuracy for the third crawl is:

$$\text{For Web Docs}: \quad AW_3 = \frac{1724}{2167} * 100 = 79.556\% \qquad \text{For pdf Docs} \quad AP_3 = \frac{387}{458} * 100 = 84.497\%$$

The plot of accuracy for web pages and pdf documents is shown in Fig. 5 for the first, second and third round of crawl. The average accuracy has been computed as follows:

For web pages, the average accuracy can be computed as:

$$AP = \frac{AP_1 + AP_2 + AP_3}{3} = 82.12\%$$

For PDF Docs, the average accuracy can be computed as:

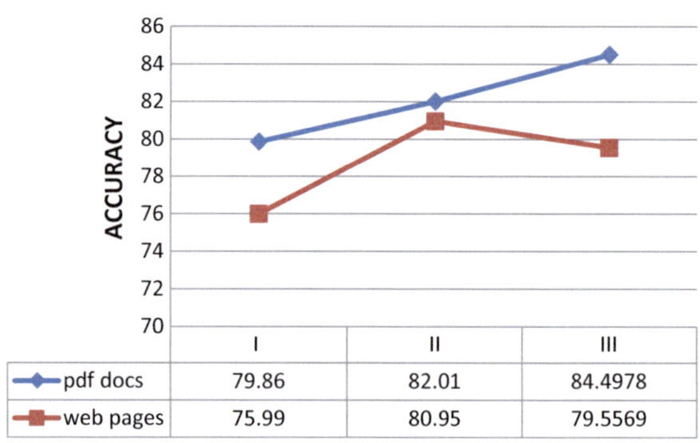

Fig. 5 Graph for accuracy of web pages and pdf documents

$$\mathrm{AP} = \frac{\mathrm{AP}_1 + \mathrm{AP}_2 + \mathrm{AP}_3}{3} = 78.83\%$$

Thus, it can be observed that the accuracy of the proposed work ranges from 79.86 to 84.5% for pdf documents and 76–80.95% for web pages. With these visible facts, it is clear that the proposed work is able to perform the task of retrieving relevant keywords from both type of content, i.e. pdf documents and web pages with a high accuracy.

5 Conclusion

The task of an image crawler is to crawl as many images as possible from WWW. In this chapter, architecture of an image crawler has been proposed that aims to target not only the surface web images but also the image in the pdf documents to be indexed. The AIC is an expansion of traditional image crawler. The very capability of it to crawl the images inside the pdf documents makes it different from other existing image crawlers

As the future work, Hidden Web may also be explored so as to make every piece of information indexable. Moreover, some techniques may be embodied to put a check on duplicity of contents in the digital form.

References

1. Harmandas, S.M, Dunlop M.D.: Image retrieval by hypertext links proceeding of ACMSIGIR (1997)
2. Marinheiro, R.N., Hall, W.: Expanding a Hypertext Information Retrieval System to Incorporate Multimedia Information. In Proceedings of the 31st Annual Hawaii International Conference on System Sciences, Vol. II, (1998)
3. Mukherjea, S., Hirata, K., Hara, Y.: Towards a Multimedia World-Wide Web Information Retrieval Engine. In Proceedings of the Sixth International World-Wide Web Conference (1997)
4. Jacso, P.: Tools for unearthing pdf files, Information Today, 48–49 (2001)
5. Kherfi M.L., Brahmi D., Ziou D., Bernardi A.: Atlas WISE: A Web based Image Retrieval Engine (2003)
6. Cho, J., Mukherjea, S.: Crawling Images on the Web, In Proceedings of Third International Conference on Visual Information Systems (Visual99), Amsterdam, The Netherlands, (1999)
7. Kherfi M.L., Bernardi, A.: What is behind image retrieval from the WWW
8. Kherfi M.L., Brahmi D., Ziou D.: Image Retrieval from the World Wide Web: Issues, Techniques, and Systems. ACM Comput. Surv. **36**(1), 35–67 (2004)
9. Remco, C.: Veltkamp, Mirela Tanase: Content-Based Image Retrieval Systems: A Survey (2000)
10. http://www.openarchives.org
11. Sharma, S., Sharma, A.K., Gupta, J.P.: Exploring OAI-PMH: Open Archives Initiative Protocol for Metedata Harvesting, IJARCS (2010)

12. Nelson M.L., Van de Sompel, H., Liu, X., Harrison, T., McFarland, N.: mod_oai: An Apache Module for Metadata Harvesting, In Proceedings of ECDL (2005)
13. Mukherjea, S., Hirata, K., Hara, Y.: Towards a Multimedia World-Wide Web Information Retrieval Engine. In Proceedings of the Sixth International World-Wide Web Conference (1997)
14. Jacso, P.: Tools for unearthing pdf files, Information Today, 48–49 (2001)

A Novel Web Page Change Detection Technique for Migrating Crawlers

Ashlesha Gupta, Ashutosh Dixit and A.K. Sharma

Abstract Change in content of web documents is a constant process and this rate of change is different for different pages. This change must be updated at the search engine database else a user gets a superseded image of the web documents. Many methods for change detection have been developed that use tree based comparisons to decide whether two versions of a web document are same or not. But these methods are prone to high complexity and ambiguity. Also frequent crawler revisits results in increased pressure on Internet traffic and bandwidth usage. In this paper network efficient web page change detection technique for migrating crawlers is being proposed that effectively detects structural and content changes by comparing proposed tag and text code for each of the html tags contained in the web page. The proposed method performs well and is able to detect changes even at minute level while keeping the network load low.

Keywords Crawler · Change detection · Network-efficiency · Migrants

1 Introduction

The World Wide Web is a heterogeneous warehouse of text documents, metaphors, and other similar items referred to as information resources [1–4]. World Wide Web has changed from a static source into a rich dynamic information delivery channel where new pages are being supplemented, modified and deleted. If this change is not mirrored at the search engine side a user may get an old image of the web

A. Gupta (✉) · A. Dixit
Department of Computer Engineering, YMCAUST, Faridabad, India
e-mail: gupta_ashlesha@yahoo.co.in

A. Dixit
e-mail: dixit_ashutosh@rediffmail.com

A.K. Sharma
Department of Computer Engineering, BSAITM, Faridabad, India
e-mail: asholkale2@rediffmail.com

S. Urooj and J. Virmani (eds.), *Sensors and Image Processing*, Advances in Intelligent Systems and Computing 651, https://doi.org/10.1007/978-981-10-6614-6_5

documents. Many methods for change detection have been developed to decide whether two versions of a web page have changed or not and whether there is a need to replace web page or not. They operate by comparing page parameters of old and new web pages using DOM Tree comparisons [2]. These methods of page detection however suffer from problems of complexity, efficiency and ambiguity [2, 5, 6]. Therefore there is a need to develop new web page detection techniques that require less time to detect changes and update only those elements such as table, images etc. that have undergone change instead of updating entire web page.

In this paper a novel page change detection technique for migrating crawlers is being proposed. In the proposed technique the process of change detection is carried out at web servers instead at search engine side which reduces network load caused by web crawlers. The algorithm is based on comparing novel tag code and text code of the old web page with new web page. If the amount of change between two versions of web page is less than specified threshold value than those changes are saved in a text file. The text file being substantially smaller file than the entire web page can be easily sent thereby consuming less network bandwidth and recourses. The modified information in the text file would then be used to update search engine repository. However if the degree of difference is more than the threshold value, new page would be sent and replaced with the old page in the search engine repository. The threshold value may be set depending on the frequency of change of the web page.

2 Related Work

A web page may change over a period of time. This change may be in Structure, Content, Behavior and Presentation of a web page. Page Change detection techniques are used to detect these changes and update the database of web pages at the search engine side. Many web page detection methods have been developed so far. For example Yadav et al. [2] proposed a change detection technique based on comparing simplified hierarchy trees of old and new web pages. Artail and Abi-Aad [5] proposed a page change technique for pages that can be changed to XML structure. The technique was based on performing similarity computations among same HTML tag types of two versions of a web page. The technique required lot of storage space and time for tree comparisons. S. Goel and Aggarwal [7] proposed comparing hash codes values assigned to nodes of the two trees for change detection. All these page detection techniques relied on constructing document trees for the web pages and were complex and required lot of space and time for tree storage and tree comparisons. These Web page detection algorithms use numerical code schemes based on ASCII values to detect content changes, they are not able to detect content changes properly when position of some of words changes. Therefore there is a need to remove this ambiguity and find accurate content changes [8–11]. Moreover change in content of web documents is an unremitting process and the change varies from document to document [3]. When the information contained in a document changes very frequently, the crawler downloads the document as often as possible and updates it into its database so that fresh information could be maintained for the potential users. If the number of such documents is large, the crawling process becomes hopelessly slow and inefficient because it puts a tremendous pressure on the internet traffic. Therefore there is a need to reduce this traffic load. Migrating crawlers can help to cop up with such frequent document changes with minimum overload over network resources.

3 Proposed Algorithm

To cater to the shortcomings discussed above a novel page change detection technique for migrating crawlers is being proposed that effectively maps both structural and textual changes. The architecture representing various components of the proposed system is shown in Fig. 1.

The major components of the proposed system are:

- Crawler Manager: It resides at the search engine side and performs various tasks such as creation of migrating agents called migrants, delegation of migrants to the Web Server, receiving updated information from agents and updating search engine's inverted index.
- Comparator: It compares old web page with the new copy of the web page (requested by migrant), detects significant changes and signals the change to the migrant. For comparing two copies of the web page, they are first converted to html format. A table of tag code and text code for each tag in the old as well as new web page in the order of their appearance is constructed next. Tag code

Fig. 1 Architecture of proposed system

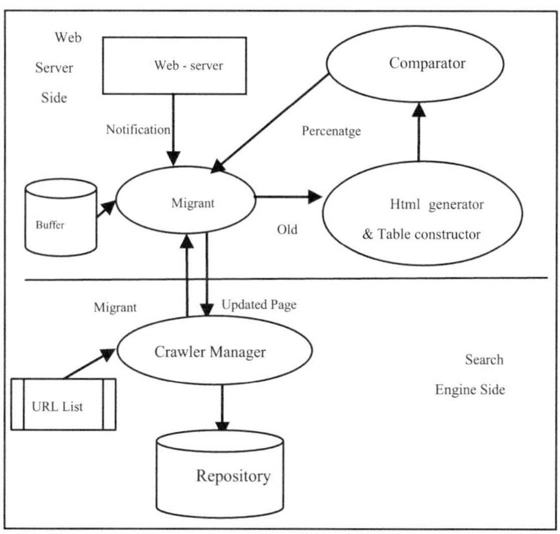

contains first two characters of each tag in the web page and the text code is created by taking a Fingerprint of the text that appears after every tag. Fingerprint is calculated based on value of alphabetic position (i.e. a = 1, e = 5, y = 25) of first and last characters of the text string and multiplying the sum of this code with length of the text string. After constructing two tables of tag and text code for the original and modified page, the two tables are compared row by row for structural as well as content changes. In order to quickly decide whether the page need to be replaced or only updated information need to be sent a threshold value for change is specified. If the page is found to be changed above the threshold value, the comparator module stops further check and signals the migrants for sending new copy of the web page to the crawler manager for storing at search engine repository. For a web page with less than threshold changes, only updated information in the form of text file is sent to the crawler. The algorithm for web page change detection is discussed in Fig. 2.

- Migrating Agent (Migrants): It is created at the search engine side and is then dispatched to reside at the web server. To keep track on the page changes at the server side migrants download/stores the web pages locally in a buffer for the purpose of duplicate page detection and filtration before sending them to search engine repository first time. After that it will keep track on modified pages. For this purpose it is expected that web server proactively inform the migrant whenever an update is made to the web page stored at that web server. After receiving notification migrant invokes comparator to find out percentage changes. If amount of change is less than specified threshold value than migrant sends a text file containing modified contents otherwise a fresh copy of the web page is sent to the crawler manager.

4 Examples

The working of the proposed algorithm is analyzed using following examples. For all the examples discussed threshold for page change is set to 30%. If amount of change is more than 30% entire page is replaced otherwise only updated information is sent back to the crawler manager.

Example 1 Consider the old and modified web pages shown in Figs. 3 and 4. The tag and text code tables for these two pages are shown in Tables 1 and 2 respectively. The two tables are compared as per the algorithm given in Fig. 2 for detecting structural and content changes.

In the above example, it can be observed that two pages differ more than 30% in content and the structure therefore new page would be downloaded and replaced with the old page in the repository.

Example 2 Consider another set of web page and its modified version as shown in Figs. 5 and 6 respectively.

1. Construct table containing tag and text value for old web page and call it otable
2. Construct table containing tag and text value for new web page and call it ntable
3. Compare rows of otable and ntable. If same proceed else goto step 12
4. set count= no of rows in otable * threshold percentage of page change allowed
5. Initialize k to 0
6. until k = count; if k=count , signal complete page change to migrant.
7. for(i=0 to last row in otable) & for (j=0 to last row in ntable)
8. compare tagvalue of ith row in otable with tagvalue of j^{th}row in ntable9. if same compare textvalue of ith row in otable with textvalue of j^{th}row in ntable else goto step 11
10. if different add to text file " value of i.tagvalue changed to j.textvalue" and increment k by 1
11. increment i , j and goto step 6
12. add to text file " i.tagvalue changed to j. tagvalue and content is j.textvalue" . Increment k by 1 and goto step 10.
13. calculate diff= (larger row value- smaller row value) / number of rows in otable *100
14. if diff >=threshold percentage of page change allowed then signal complete page change to migrant.
15. else compare rows in otable and rows in ntable
16. If rows in otable< rows in ntable proceed else goto step 9
17. for i=0 to lastrow in old table and j=0 to lastrow in new table
18. let a and b be two string variables . Set a=i.indexvalue and b=j.indexvalue
19. if (a!=b) then proceed else goto 8
20. if lengthof(a) < t lenghtof(b) then proceed else goto 24
21. shift ith row below and add an empty row above it and call newly allocated row as ith row.
22. copy contents of jth row of ntable to this new ith row.
23. add to text file " node j.tagvalue added to new page "
24. increment i and j and goto step 17
25. shift jth row in ntable below and add an empty row above it and call newly allocated row as jth row.
26. copy contents of ith row of otable to this new jth row.
27. add to text file " node i.tagvalue deleted from old table"
28. increment i and j and goto step 17

Fig. 2 Algorithm for the proposed system

Fig. 3 Old web page

<html>
<head> YMCA Page </head>
<body>
<p> Important News</p>
<p> M.Tech Councelling from 23/7/15</p>
<p> Applicants should deposit forms<b> Latest by 15/7/15</b>
</p>
</body>
</html>

```
<html>
<head> YMCA Home Page </head>
<body>
<p> Latest Alerts</p>
<p> Phd. Councelling from 31/7/15</p>
<p> Applicants should deposit
forms<u> Latest by 22/7/15</u>
</p>
</body>
</html>
```

Fig. 4 Modified web page

Table 1 Tag and text code table for old web

Html tag	Tagcode	Contents	Text code
<html>	<ht>		
<head>	<hthe>	YMCA page	270
</head>	<ht>		
<body>	<htbo>		
<p>	<htbop$>	Important news	392
</p>	<htbo>		
<p>	<htbop$>	M.Tech councelling from 23/7/15	378
</p>	<htbo>		
<p>	<htbop$>	Appicants should deposit forms	728
<b>	<htbop$b$>	Latest by 15/7/15	153
</b>	<htbop$>		
</p>	<htbo>		
</body>	<ht>		
</html>	<>		

Tag and text code comparisons for the above two pages using the proposed algorithm were performed and it was observed that two pages differ in content and change is less than 30%. So a text file with information about changed content would be passed to the crawler manager. Based on above examples it can be concluded that the proposed method works well with different pages of different and varied formatting styles accurately capturing their structures at any given point of time.

A comparative study of the proposed page change detection algorithm using Migrants and Web page change detection based on simplified hierarchy tree of HTML Document using Parallel crawler architecture was also conducted on various factors. The results of the study are summarized in Table 3.

Table 2 Tag and text code table for modified page

Html tag	Tagcode	Contents	Text code
<html>	<ht>		
<title>	<htti>	YMCA Home page	420
</title>	<ht>		
<body>	<htbo>		
<p>	<htbop$>	Latest alerts	403
</p>	<htbo>		
<p>	<htbop$>	Ph.D. councelling from 31/7/15	441
</p>	<htbo>		
<p>	<htbop$>	Appicants should deposit forms	728
<u>	<htbop$u$>	Latest by 23/7/15	187
</b>	<htbop$>		
</p>	<htbo>		
</body>	<ht>		
</html>	<>		

Fig. 5 Original web page

```
<html>
<head> News <./head>
<body>
<p> Latest News : Dr. APJ Abdul Kalam pass-
es away</p>
< a href="www.timesofindia.com"> Click for
complete info.</a>
</body>
</html>
```

Fig. 6 Original web page

```
<html>
<head> News <./head>
<body>
<p> Latest News : President Pranab
Mukherjees wife passes away </p>
<a href="www.timesofindia.com"> Click for
complete info.</a>
</body></html>
```

Table 3 Comparison between proposed technique & simplified hierarchy tree based scheme

S.No	A novel page change detection algorithm using migrants	Parallel crawler architecture and web page change detection based on simplified hierarchy tree of HTML document
Technique for detection	Table rows comparison	Tree comparison
Type of changes detected	Structural & content	Structural & content
Can detect both types of changes simultaneously	YES	NO
Methods required to detect changes	Single method to detect both types of changes	Two methods: one for structural and one to detect content
Complexity	Less	High
Time to detect & replace page	Fast	Medium
Time complexity	Linear O(n)	Linear O(n)
Memory requirement	Less	Large

5 Conclusion and Future Work

In this paper a new algorithm for detecting structural and content changes in web pages is proposed and described. The proposed algorithm is fast and does not require scanning of entire web page to decide for page replacement. As soon as the specified amount of change is detected, the algorithm automatically signals page replacement.. The proposed system handles the shortcomings of centralized crawling system and is more robust to significant web page changes than other existing change detection techniques that depend on page size, hash value and tree based comparisons. The use of migrants helps in reducing network load significantly thereby saving memory and bandwidth overheads. As a future work the algorithm needs to be extended for behavioral changes and changes in images or other media files.

References

1. Chakravarthy, Hari H.S.C.: Automating change detection and notification of web pages. Proce. 17th Int. Conf. Database Expert Syst. Appl. (DEXA'06), IEEE, 0-7695-2641-1/06, (2006)
2. Yadav D., Sharma A.K, Gupta J.P: Topical web crawling using weighted anchor text and web page change detection techniques. 10th Int. Conf. Inform. Technol. IEEE, 0-7695-3068-0/07, 265–270, (2007)
3. Sharma AK, Dixit A.: Self-adjusting refresh time based architecture for incremental web crawler. Int J Comput Sci Network Secur(IJCSNS), **8**(12), 349–54, (2008)

4. Gupta Ashlesha, Dixit Ashutosh.: Issues and Challenges in Effective Design of Search Engine. Int. J. Multi. Res. Studies, Dec (2012)
5. Artail H. and Abi-Aad M: An enhanced web page change detection approach based on limiting similarity computations to elements of same type, Springer Science + Business Media. LLC. pp. 1–21 (2007)
6. Yadav D., Sharma A.K.,Gupta J.P.: Parallel crawler architecture and web page change detection. WSEAS Trans. Comput. pp 929–940, (July 2008)
7. Goel S., Aggarwal R. R.: An efficient algorithm for web page change detection. Int. J. Comput. Appl. (0975—888), **48**(10), 28–33, June (2012)
8. Wang Y., DeWitt D, Cai,J.: X-Diff: An Effective Change Detection Algorithm for XML Documents. Proc. 19th Int. Conf. Data Eng. pp. 519–30, (2003)
9. L. Su-bin, W.C. Shi, Z.H Liang, X.M.Yu, L. Zhang.: A direct web page templates detection method. IEEE Int. Conf. 978–1-4244-7255-0/11, (2011)
10. P. Ying, D. Xuhua.: Anomaly based web phishing page detection. Proc. 22nd Annu. Comput. Secur. Appl. Conf. IEEE, 0-7695-2716-7/06, (2006)
11. G. Ashlesha, Dixit, A., Sharma A.K.: Relevant document crawling with usage pattern and domain profile based page ranking. ISCON, 2013, IEEE International Conference held at GLA University, Mathura, (2012)

Developing Prototype for Prosopagnosia Using PCA

Gunjan Jhawar, Prajacta Nagraj and N. Ramesh Babu

Abstract Prosopagnosia primarily known as face blindness is a brain disorder in which a person is not able to recognize faces. In the device a portable camera is used to capture the real time movement of a person. In the main system face is detected using Voila –Jones Algorithm & hue channel detection. Using Principal Component Analysis (PCA) face is recognized. It is compared with the trained database and displays the basic information about the person if a similar trained data is found otherwise, it is declared as unknown. For new database, system can be trained at any point of time. The prototype system is trained using 30 candidates' images including 10 images per candidate under distinct angles. The system was able to achieve 98% face detection. It provides the patients an aid to help them recognize people in front of them in seconds.

Keywords Face recognition · Face detection · Principal component analysis · Eigen values · Viola jones algorithm

1 Introduction

Prosopagnosia is a neurological disorder where the ability to recognize faces is impaired. It is also known as face agnosia. It is often accompanied by other types of recognition impairments (place recognition, car recognition, facial expression of emotion etc.) though sometimes it appears to be restricted to facial identity though characteristics like intellectual functions remain intact [1].

G. Jhawar · P. Nagraj · N. Ramesh Babu (✉)
School of Electrical Engineering, VIT University, Vellore, Tamilnadu, India
e-mail: nrameshbabu@vit.ac.in

G. Jhawar
e-mail: jhawargunjan@gmail.com

P. Nagraj
e-mail: nagraj.prajacta@gmail.com

S. Urooj and J. Virmani (eds.), *Sensors and Image Processing*, Advances in Intelligent Systems and Computing 651, https://doi.org/10.1007/978-981-10-6614-6_6

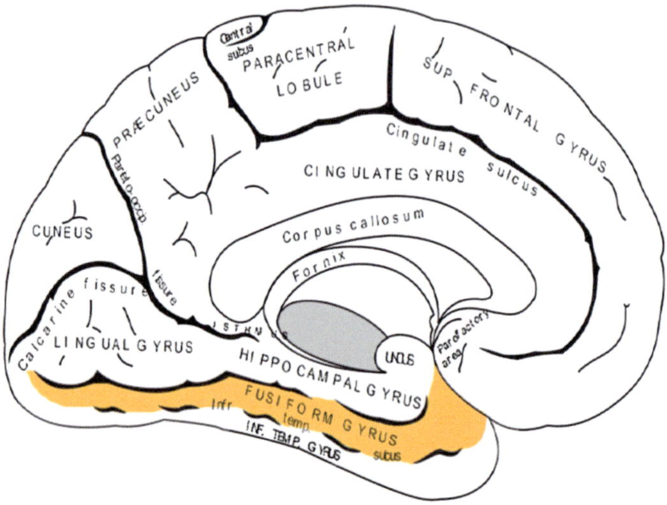

Fig. 1 Cross section of brain showing fusiform gyrus by *gray*, vectorized by mysid, colourd by was_a_bee released to public domain. [4]

The specific brain area usually associated with prosopagnosia is the fusiform gyrus. The functionality of the fusiform gyrus allows most people to recognize faces in more detail than they do similarly complex inanimate objects [2]. Figure 1 shows the cross-section of the brain image with fusiform gyrus. Figure 2 shows the pictorial representation of state of the person affected by prosopagnosia.

There are two types of prosopagnosia: acquired and congenital (developmental). Acquired prosopagnosia results from occipitotemporal lobe damage and occurs due to brain damage from trauma, stroke and degenerative diseases. Developmental prosopagnosia occurs since birth. A genetic factor is responsible for this condition [3].

Existing devices used for aiding people suffering from above disorder includes Google glass which helps people recognize various faces. However this technology uses online profiles of various users from social networking site which in turn puts this device in jeopardy caused by risking privacy of other people. This device functions offline which eliminates the risk by training images of people selected by the person suffering from prosopagnosia.

2 Methodology

The prototype system proposed is made up of four steps, at the end of which the person using the device would be able to get the information related to the face in the input video. The flowchart of the proposed system is clearly shown in Fig. 3. The first step is capturing the video of the person. It is being done with the help of

Fig. 2 "Prosopagnosia, the inability to recognize faces" by krisse used under creative commons attribution- share alike 3,0 unported license [5]

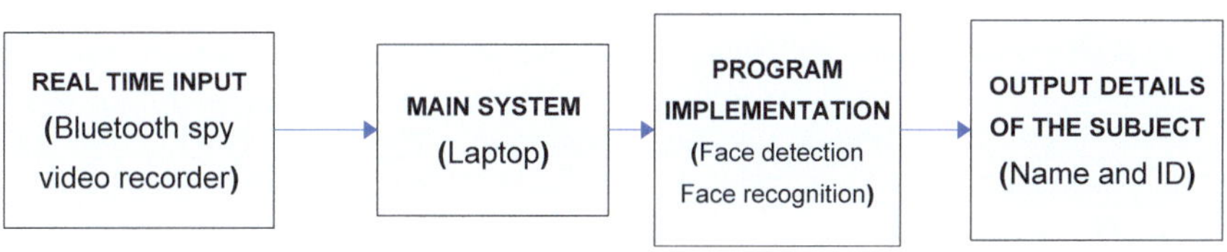

Fig. 3 Flowchart depicting the basic working of the device

bluetooth camera which captures the video for 20 s. This real time video is captured in format of MP4 (we used camera with specs 16 frames/sec, frame size of 640*480 and length of 12 s).

Second step involves converting the video to AVI format after which a single frame of video is extracted for face detection. After the video file is received face detection is carried out using MATLAB as follows.

2.1 *Face Detection*

Firstly, MP4 video (16 frames per sec) is converted to AVI index for further processing, for detecting a face from live video using vision, Cascade Object Detector. The cascade object detector uses the Viola-Jones detection algorithm and a trained classification model for detection.

This detector is valid only in case of constant posture, if the person tilts or faces sidewise, upwards, downwards it fails to detect the face. For this purpose we use a unique feature to differentiate face from other objects which is skin tone (hue channel data), which corrects the glitches faced by former method.

2.2 Face Recognition

Second step and most important one is to recognize face of a person and this is done using MATLAB. Firstly, a database is trained in the model according to the user (relatives, friends etc.) acquaintance, and then the required image is collected from above face detection technique whose information is needed. Using PCA algorithm we determine the closest face present in trained model and display its information accordingly. The following flow chart explains how the algorithm works [6–8].

1. Select New Image: It selects an image used for training the database or to recognize a face.
2. New Record: It trains the new selected Image and forms a new record storing Details of the person (Name, ID no. etc.)
3. Number of ID's: Displays Individual person's information for the given ID number.
4. Face Recognition: PCA extracts the suitable information in a face image and encodes as efficiently as possible. It identifies the subspace of the image space covered by the training face image data and de correlates the pixel values. The representation of a face image is obtained by projecting it to the principal components. The projection of face images into the principal component subspace achieves information compression, decorrelation and dimensionality reduction to facilitate decision making (Fig. 4).

In mathematical terms, the principal component of the distribution of faces or the eigenvectors of the covariance matrix of the set of image is searched by treating an image as a vector.

Algorithm for PCA is described as follows:

1. Subtract the Mean of the data from each variable (our adjusted data): It is the difference of the mean of each row from each element in that row.
2. Calculate and form a covariance Matrix:
   The basic Covariance equation for two dimensional data is:

$$\operatorname{cov}(x,y) = \frac{\sum_{i-1}^{n}(x_i-\bar{x})(y_i-\bar{y})}{n-1} \tag{1}$$

3. Calculate Eigen vectors and Eigen values from the covariance Matrix:

Eigen values are a product of multiplying matrices and are found by multiples of the covariance matrix by a vector in 2 dimensional space (a Eigenvector)
4. Chose a Feature Vector (a fancy name for a matrix of vectors)
5. Multiply the transposed Feature Vectors by the transposed adjusted data.

3 Testing

The system was tested using 5 individuals, 10 images per individual in the database. It was thereafter linked with the video which consisted of two individuals out of these five as shown in Fig. 5. In which one of them is sitting in front of the camera and then the other one is walking behind (with noise). The developed prototype system successfully identified both the subjects after detection from real time video.

4 Conclusion

Result: The system was able to recognize the person sitting in proximity and compare it with the trained database. It matched the image of the input to the stored database with an accuracy of 98%.

Usage: This device can be used by people suffering from face blindness, also its application is not limited to this context. It can be extended to blind people where in the information output is audio. For mobility this device can be extended using MATLAB mobile, which connects with the laptop using cloud.

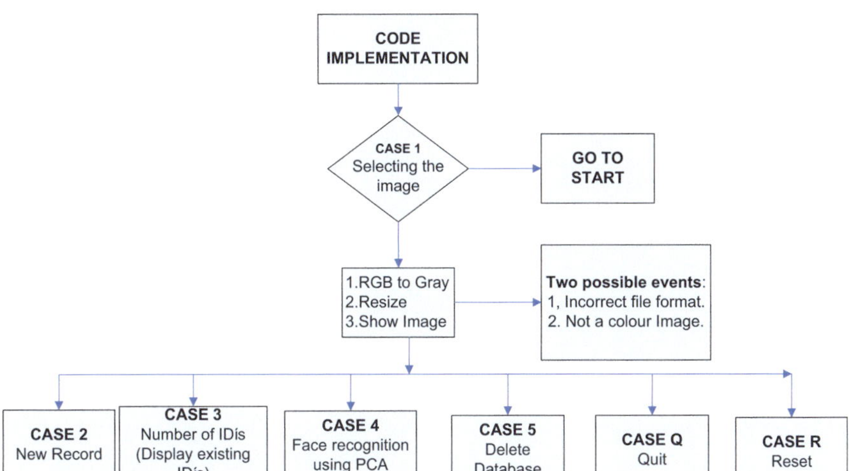

Fig. 4 Depicting the working of face recognition

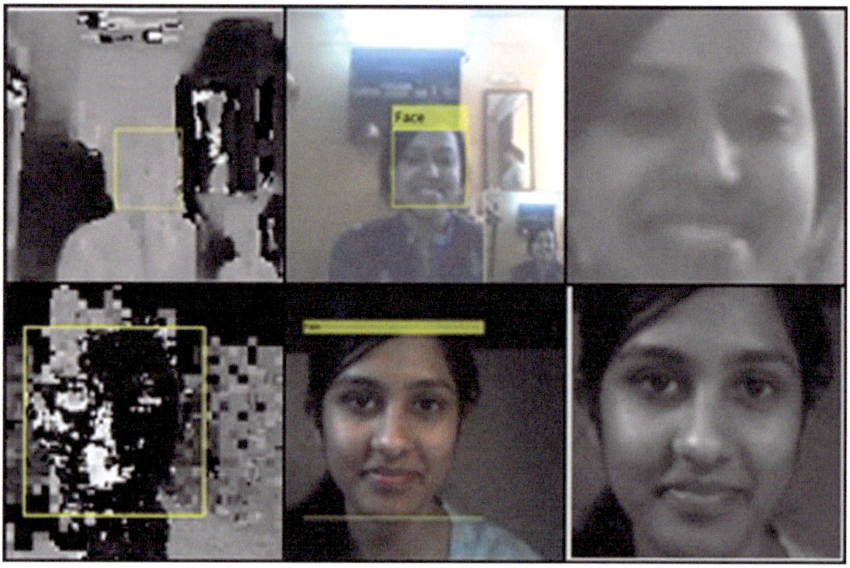

Fig. 5 Images showing hue channel detected face, detected face from video and final cropped face ready for recognition respectively

References

1. https://www.faceblind.org/research/.
2. Vandermeulen, R., Morissette, L., Chartier, S.: Modeling prosopagnosia using dynamic artificial neural networks. Neural Networks (IJCNN), The 2011 International Joint Conference on, pp. 2074, 2079 (2011)
3. https://en.wikipedia.org/wiki/Prosopagnosia#cite_note-cnn2407-3.
4. http://www.google.co.in/imgres?imgurl=http://www.todayifoundout.com/wpcontent/uploads/2012/12/fusiformgyrus.
5. https://commons.wikimedia.org/wiki/File:Prosopagnosia.jpg.
6. G.R. Bradski. Real time face and object tracking as a component of a perceptual user interface. proceedings of the 4th IEEE workshop on applications of computer vision, (1998)
7. Viola, Paul A., Jones, Michael J. Rapid object detection using a boosted cascade of simple features. IEEE CVPR (2001)
8. http://www.mathworks.com/matlabcentral/fileexchange/loadFile.do?objectId=6995&objectType=file.

Implementation of Smart Video Surveillance System Using Motion Detection Technique

Gaurav Verma, Swati Gautam, Rohan Agarwal, Sarthak Saxena and Deval Verma

Abstract In today's world, there is a growing need of protection of one's property—security is one major issue which should be taken care of by everyone. There is a lot of increase in the number of thefts or stealing with the advancement of technology and so this directs our attention toward protecting our property. The most popular way of protecting our property would be a video surveillance and constant monitoring. Since technology has both boon and harm, we can use it to develop a system to improve the functionalities of existing video surveillance system, which can track the movements of the suspicious people, generate an alarm, and even record them from the time of suspicious activities. Our main motive is to develop a real-time application for the purpose of video surveillance which will not only detect motion, but also inform the user of any kind of suspicious activity, record from the moment of any kind of activity, and also turn the camera in the direction of motion. We will be doing it with the help of a MATLAB version 2014a.

Keywords Video surveillance · Background subtraction · Blob analysis · Arduino · Motor rotation

---

G. Verma (✉) · S. Gautam · R. Agarwal · S. Saxena · D. Verma
Electronics and Communication Engineering, Jaypee Institute of Information Technology, Noida, Uttar Pradesh, India
e-mail: gaurav.iitkg@gmail.com

S. Gautam
e-mail: gswati025@gmail.com

R. Agarwal
e-mail: 23rohan.agarwal@gmail.com

S. Saxena
e-mail: ssarthak.9424@gmail.com

D. Verma
e-mail: deval09msc@gmail.com

S. Urooj and J. Virmani (eds.), *Sensors and Image Processing*, Advances in Intelligent Systems and Computing 651, https://doi.org/10.1007/978-981-10-6614-6_7

1 Introduction

Representation of an image can be in the form of two-dimensional function and can be written as f(x,y), i.e., the function of (x,y) where x and y are coordinates and magnitude/amplitude of this point is called intensity/gray level of image at these coordinates. When the magnitude values of function f are finite implying discrete values, the image hence formed is a digital image [1]. Digital Image Processing is the manipulation of digital images by the use of computer algorithms. Digital Image Processing is better than analog image processing as it allows a much wider range of computer algorithms to be applied to the input image. Digital Image Processing is a subset of signals and systems and emphasizes only on images [4, 5]. In order to convert analog signals to digital images, we require the use of various computer algorithms. Digital Image Processing is based on the use of such algorithms. They improve clarity, image compression ability across networks for communicating purposes, size extraction, etc., [3]. Digital images can be enhanced in many ways, for example, we can increase the brightness, convert the image to black and white and vice versa, even the colors can be changed for further processing [2]. MATLAB helps us to store images as two-dimensional arrays or matrices. Each element of the matrix represents one pixel in the image. Let us say we have an image of 500 rows and 500 columns. This 500 x 500 will represent a matrix having 10,000 different colored dots or pixels to be more precise. A pixel is combination of RGB, i.e., red, green, and blue components. But in case of RGB, the first plane in the third dimension represents red pixel depth, the second plane is for green pixel depth, and the third plane is for blue pixel depth or intensities [3]. Motion detection is one of the most popular areas which help in developing a security system. The most popular technique used in image processing and computer vision is background subtraction. In this process, we extract the foreground of the image required for various purposes during processing. And hence, it is also known as foreground detection. Since the moving objects like human bodies in the foreground of images are the main things of our concern, background subtraction is widely used approach in "motion detecting" field. The main principle behind such technique is subtracting the current frame from previous frame [6]. The organization of this paper is such that we have an introduction in Sect. I, then we have objective of the paper in Sect. II, system model in Sect. III, acquisition setup in Sect. IV, image processing in Sect. V, experimental results in Sect. VI, applications in Sect. VII, and conclusion in Sect. VIII.

2 Objective

The requirement of a smart video surveillance system provoked us to develop this project which will capture any kind of motion and thus will help us prevent any kinds of thefts or stealing at the very initial stage and will also provide us evidence in the form of video recording. We will be well prepared for future by installing this system [2].

3 System Model and Algorithm

MATLAB (MATrix LABoratory) is used for several mathematical operations and finds its use worldwide. It has the ability to solve a lot many problems related to DSP (digital signal processing) or image processing, it has a vast use in communications, used for various educational purposes, research work, and even industrial work. It has user-friendly interface and is developed by MathWorks. MATLAB provides variety of toolboxes for image handling purposes. The toolboxes involved were image acquisition toolbox (for grabbing the image), image processing toolbox, and computer vision system toolbox (for blob analysis).

The algorithm is as follows:

- Start capturing the video.
- Initialize the RGB components of the first frame as zero. Store the components in a temporary matrix.
- Turn the image by 180 degrees. For this purpose, we can use the flip functions. (We are doing this because we need to store the correct image and not the mirror image.)
- Compare two successive frames (means comparing the RGB components of each frame → matrices). Perform background subtraction.
- If motion produced (means if there is nonzero value achieved after subtraction), then recording starts, alarm beeps.
- At the same time, performing median filtering to highlight the major patches produced during subtraction. Median filtering is required to smooth the edges, highlight them, and keep the major patches or "blobs" with which motion is concerned in the differenced image produced after image subtraction.
- RGB to grayscale conversion: A pixel in the 8-bit grayscale image has magnitude or intensity values ordering from 0 to 255. This means a pixel can have any of these 256 values. People confuse gray scale with black and white image. In actual, a gray scale is an image which contains many shades of gray. Since the conversion of RGB to gray scale is as good as moving from 3-D operations to 1-D operations, therefore in order to reduce our load of work in dealing with the three components of RGB, we switch to gray scale, i.e., from an intensity of 24 bit (8×3) to an intensity of 8 bit.
- Gray scale to binary image: For converting, we use im2bw( ) function. Suppose we have set a level or threshold, then all the pixels whose intensity values are above that level will be treated as value 1 and all other pixels will be replaced by value 0. Zero corresponds to white color and one corresponds to black color. Thus in this case, further values are reduced to two—0 and 1.

- Blob analysis to find the centroid of the blobs and store the centroids in a matrix.
- Centroid subtraction.
- Pass the differenced pixel values or say manipulated values to Arduino.
- Motor rotation (which in turn rotates camera).

4 Acquisition Setup

For acquiring image/video, we require image acquisition toolbox. With the help of this toolbox, we can detect our hardware easily and set properties of the hardware according to our requirements. The basic steps are as follows:

1. Accessing the webcam's video input.
2. Acquiring the image and storing it in the form of a variable in MATLAB.
3. Using the acquired image for further processing [7].

[8] First step for connecting the image acquisition device in MATLAB is creating a video input object. The video input object establishes the connection between image acquisition device and MATLAB. There are several object-related properties that can be used for the purpose of acquisition and controlling its various aspects. A proper procedure has to be followed before creating the object. First of all, we need details about the device to which we are establishing a connection. In order to access the device, the image acquisition toolbox should have details about the name of adaptor (adaptor is the interface which is required for establishing connection between image acquisition devices connected to the system and MATLAB), the device ID, video format, etc. So, an important step is to acquire information about our available hardware on MATLAB. For this purpose, we use the "imaqhwinfo" function which returns information on available hardware. This comes under the image acquisition toolbox. It should be noted that "Start" function should be used to start the object before starting acquisition of data.

5 Image Processing

The toolboxes which are used after acquiring the image are image processing toolbox & computer vision systems toolbox having following features: Image analysis including segmentation, image enhancement, filtering, and de-blurring, image transforms, including FFT etc. Provides functions such as "rgb2gray", "imsubtract", "imread", "imwrite" etc. [9–12]. Computer vision system toolbox supports large videos or handles large videos. It helps us to estimate motion using various functions such as BlobAnalysis, VideoPlayer, and ShapeInserter. We required it for various purposes—blob analysis, recording video functions.

6 Experimental Results

The following are the results observed at the end of the experiment. We took four screen shots—for moving the camera in the downward, left, right, and upward direction. For example, when the moving body reached the rightmost edge of the frame, manipulations of the centroid sent the command to the Arduino, which made the motor move in the right direction (Figs. 1 and 2).

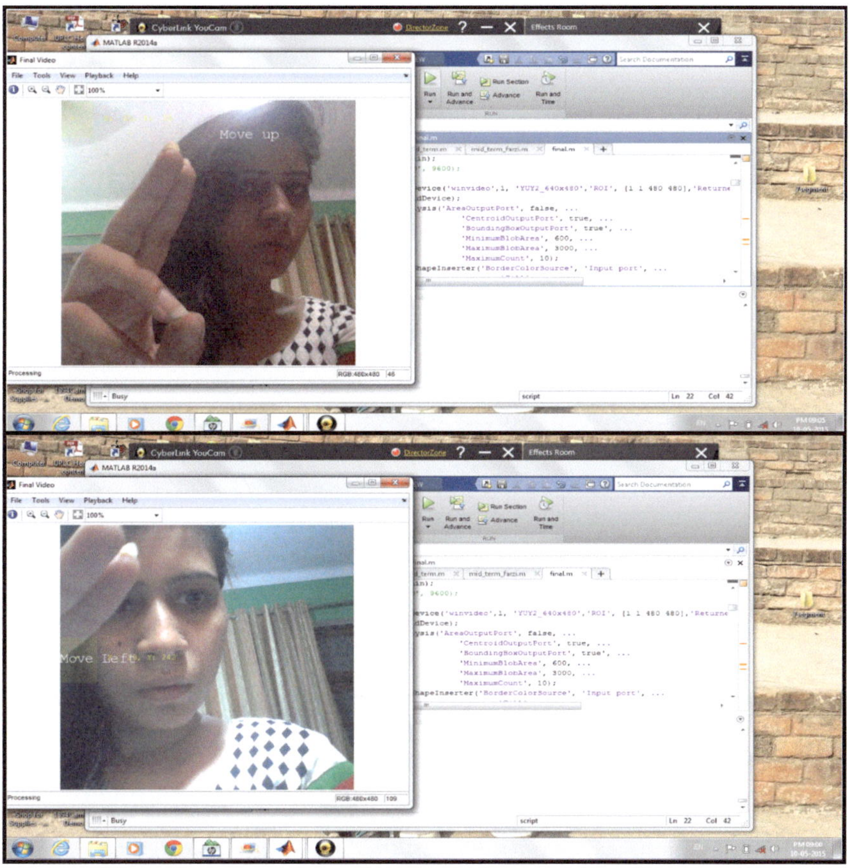

Fig. 1 Showing camera movement in up and left direction

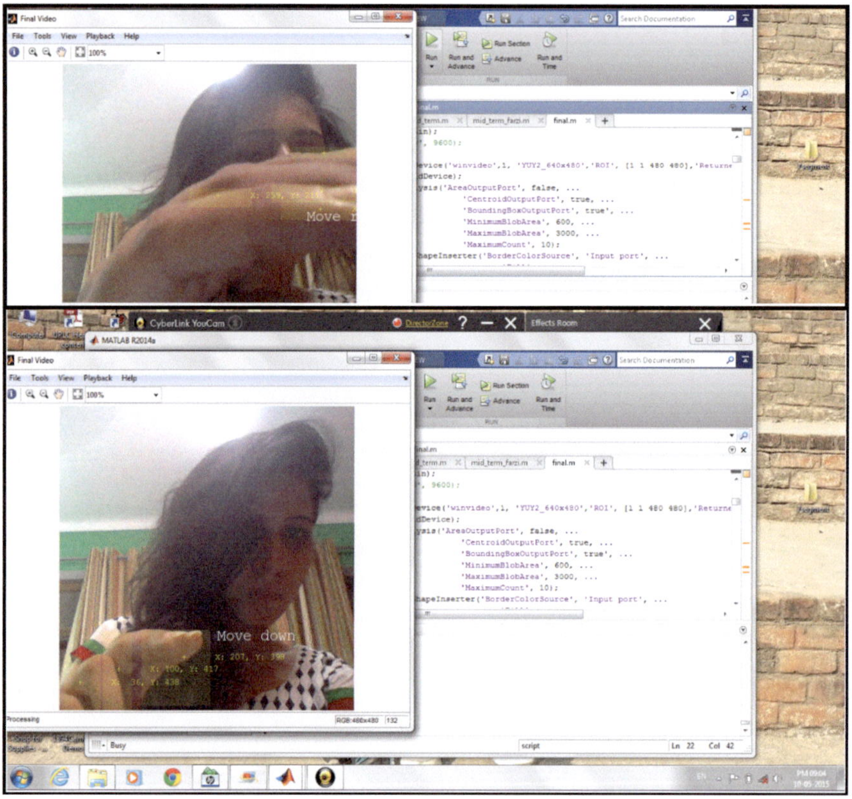

Fig. 2 Showing camera movement in down and right direction

7 Applications

Our main objective is developing this smart surveillance system which will not only detect motion, but also move the camera in the direction of motion. Also, if the motion is in the restricted area or if there is mishandling with the camera, then generate an alarm informing the user of any suspicious activity and also record the video for future purposes meant for processing. This can be used in motion sensing games. It can be of great help to the camera man in cricket matches. Since in this project, we are moving the camera according to the motion, it can reduce the manual work of the camera man to follow the ball with such heavy cameras. We can do this by monitoring the camera man's motion to follow the ball. It increases the range of area of camera to be observed (since the camera is rotating).

8 Conclusion

We were able to build a video surveillance system for detecting the motion and thus preventing or lessening the amount of thefts or stealing and thus providing a sense of security to people or any organization and also providing the proof of any mishappening in the form of mp4 extension file. The work can be further extended to provide more additional functionalities. These are as follows: The video can be transferred to the Internet to be viewed by everyone on the cloud servers. This underlines the concept of "INTERNET OF THINGS." The details can be viewed on the smart phones as well. A log-in system can be implemented so that the software can be used only by the authorized people. A proper GUI can be prepared for the layman user so that there is no problem in understanding as to how to run the application. Inbuilt audio recorders, so as to record sound too can be added.

9 Consent

We don't have any objection to publishing this paper as a chapter with the images shown above. As the images used are of one of the co-authors of this paper Ms. Swati Gautam.

References

1. http://elearning.vtu.ac.in/17/e-Notes/DIP/Unit1-SH.pdf
2. A. Ansari, T.C. Manjunath (Ph.D., IIT Bombay) and C.Ardil, Implementation of a Motion Detection System, International Journal of Electrical, Computer, Electronics and Communication Engineering **2** 9, (2008)
3. www.mathworks.com
4. http://www.tutorialspoint.com/dip/
5. Suresh D.S and Lavanya M.P.: Motion Detection and Tracking using Background Subtraction and Consecutive Frames Difference Method, International Journal of Research Studies in Science, Engineering and Technology, **1**(5), 16–22, ISSN 2349–4751 (Print) & ISSN 2349-476X (Online) August (2014)
6. Jeevarajan, Matlab Code for Background Subtraction: https://www.pantechsolutions.net/blog/matlab-code-for-background-subtraction/#Posted on December 18, 2013
7. Image Acquisition Toolbox User's Guide by Mathworks, MATLAB
8. Getting Hardware Information: www.mathworks.com, © 1994–2015 The MathWorks, Inc
9. Image Processing Toolbox User's Guide by MathWorks
10. Introduction to image processing by ESA/Hubble: https://www.spacetelescope.org/static/projects/fits_liberator/image_processing.pdf.

11. T. Gupta, G. Verma: "Area & Power Optimization of VPB Peripheral Memory for ARM7TDMI Based Microcontrollers" in International Conference on Cognitive Computing and Information Processing (CCIP-2015) March 3–4, 2015 JSSATEN, Noida, India
12. G. Verma, V. Verma, S. Jhambhulkar, H. Verma, Presented a Paper "Design of a Lead-Lag Compensator for Position Loop Control of a Gimballed Payload" in International Conference on Signal Processing and Integrated Networks (SPIN-2015) February 19–20, 2015 Amity University, Noida, India

Intelligent Algorithm for Automatic Multistoried Parking System Using Image Processing with Vehicle Tracking and Monitoring from Different Locations in the Building

Sagar Juneja, Saurav Kochar and Sachin Dhiman

Abstract This paper presents an intelligent algorithm implemented to design an intelligent parking system in an existing multistoried parking facility. The system will automatically guide the user (driver) to vacant parking slot; the system will keep the track of vehicles by remembering the vehicle number and will allow user to locate the exact parking slot of his vehicle in the multistory parking, from any location in the building. All this is done with the help of image processing using image binarization and segmentation techniques and with very minimal use of hardware to ensure that the cost of the system is very small.

Parking is a very big problem today, especially in the big cities, and we are well aware of it. We have built multistoried parking facilities in malls, multiplexes, large crowded market places, but these parking facilities are manually operated; as a result, users face problems such as difficulty in ticketing, locating a free parking slot for parking, locating where their vehicle is parked among 1000s of vehicles. In our system, we have used two cameras and MATLAB for image processing. We have created a pseudoparking space with parking slots and obtained our results. As you will find out from the snapshots in the description below, it is shown that our system successfully (1) identifies the vehicle with its registration number, (2) identifies the available parking slots, and (3) ties the parking slot to vehicle number and stores it in the memory so that users can locate their vehicle.

Keywords Image processing · Automatic car parking · OCR · Number Plate Recognition · MATLAB

S. Juneja (✉) · S. Kochar · S. Dhiman
Department of Electronics & Communication Engineering,
Chitkara University Himachal Pradesh, Solan, India
e-mail: sagar.juneja@chitkarauniversity.edu.in

S. Kochar
e-mail: kochar1993@gmail.com

S. Dhiman
e-mail: sachindhiman1@live.in

S. Urooj and J. Virmani (eds.), *Sensors and Image Processing*, Advances in Intelligent Systems and Computing 651, https://doi.org/10.1007/978-981-10-6614-6_8

1 Introduction

India is one of the biggest auto markets with over 40 million vehicles on road already [1]. In any average-sized Indian city, almost 40% of the road spaces are used for parking a vehicle rather than the movement of the traffic [2]. This kind of system sadly causes a lot of problems such as traffic congestion, accidents, environment problems, obstruction to the emergency services. In country like ours, where sales of automobiles did not even slow down, during the recession period of 2009 and 2010 [1], there is a demand for very effective and smart parking system. Needless to say, the problem of parking is worldwide and there are numerous systems being adopted worldwide over the period of time to tackle the problem [3] like off-street parking, 30-degree parking, 45-degree parking, 90-degree parking, underground parking, multistoried parking, and advanced multilevel parking.

During our research for the subject, we came across that in India there are lot of companies that are making advanced multilevel parking systems that require a mechanical setup and incur huge cost [4], but we could not find any good, smart solution for existing multistoried parking buildings. In India, lot of malls, multiplex, shopping complex, housing complex, and busy commercial complex have multistoried parking of concrete buildings (mostly underground) and they are manually operated; as a result, users find the following difficulties—identifying the free parking slot available in the parking, locating the exact spot where they had parked the vehicle in the multistoried parking, etc. Secondly, in the existing scenario, there is a need for manual resources to be deployed to keep track of the vehicles. We have developed a system where all these processes are automated using an intelligent algorithm on MATLAB and a couple of cameras. This ensures that overall cost of the system shall remain the minimum.

2 Technical Background

Several systems exist for automated parking systems, but they are based on magnetic, ultrasonic, IR sensors [5, 6]. When we read about those systems, we figured out that they require a lot of hardware to be employed in them which adds to the cost and also they would require a lot of maintenance. The chances of failures and getting false alarms are higher in such systems. Lastly, we found out vehicle tracking was difficult to employ with these systems as it would have required us to use a separate technology. We wanted one single technology solution for our complete system to make it robust. We decided to implement our problem statement using image processing. We found out some work done using image processing as well [7], but the algorithm was more complex and they are considering the entire parking area during processing. However, in our system, we have used simpler approach where only parking slots are singled out and processing is done; secondly, our system also has provision of locating the vehicle which is a novel idea and not being implemented yet in any automated parking system. For vehicle tracking and locating, we are capturing the images of number plates of the vehicles, performing the image processing to extract registration numbers, and maintaining a log file. The related approach is implemented for a different application in [8].

3 Proposed Solution

Our system consists of three subsystems, all working simultaneously as shown in the following block diagram (Fig. 1). The first subsystem is deployed at the entrance of the parking slot, which consists of a camera that captures the image of number plate of the vehicle by Optical Character Recognition (OCR) technique and maintains a log file of registration numbers of the vehicle. The second subsystem consists of another camera in the parking slot which monitors and transmits real-time availability of parking slot information through GUI to administrator and to a display near the entrance. The display near the entrance guides the user to the nearest available parking slot. The third subsystem is a back-end system which ties the parking slot and parking floor number with the vehicle registration number along with date and time stamp and maintains a log. This enables user to locate the parking slot of his vehicle from anywhere in the building by just entering his

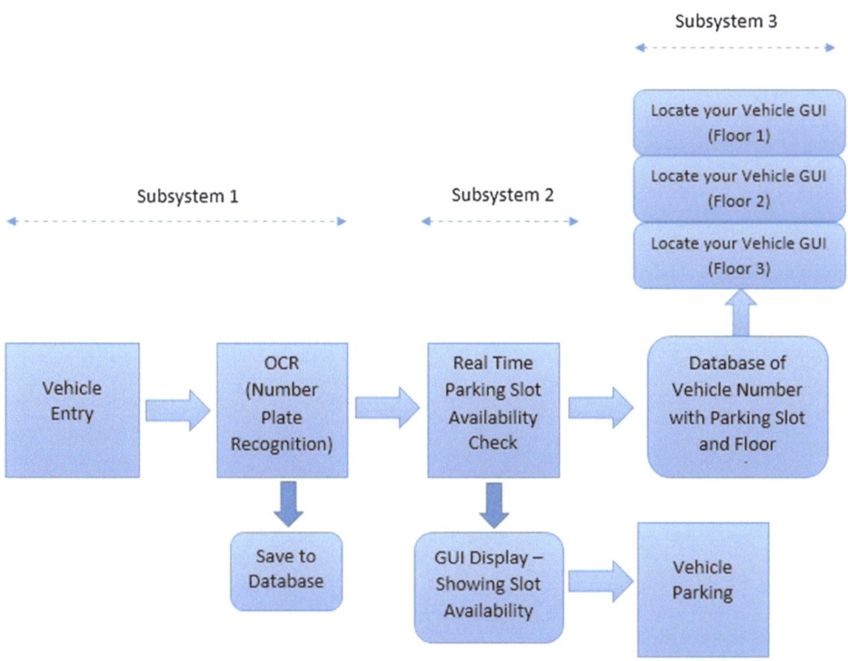

Fig. 1 Block diagram

vehicle registration number (or a part of it) in GUI developed by us. Our GUI runs on displays put up on different floors of buildings (such as malls, multiplexes).

4 Implementation

We are presenting our implementation details in two parts: software implementation and hardware implementation.

4.1 Software Implementation

In line with the system block diagram, our software implementation details have three parts.

Number Plate Recognition Figure 2 is the flowchart of Number Plate Recognition, where we have used the technique of Optical Character Recognition (OCR). The image of number plate of vehicle is obtained by the camera in RGB form. This RGB image is converted into grayscale image with the threshold of 0.5 taking in account the ambient light conditions. The image is then converted into a binary image. Then, we have performed preprocessing on the binary image to

Fig. 2 Flowchart

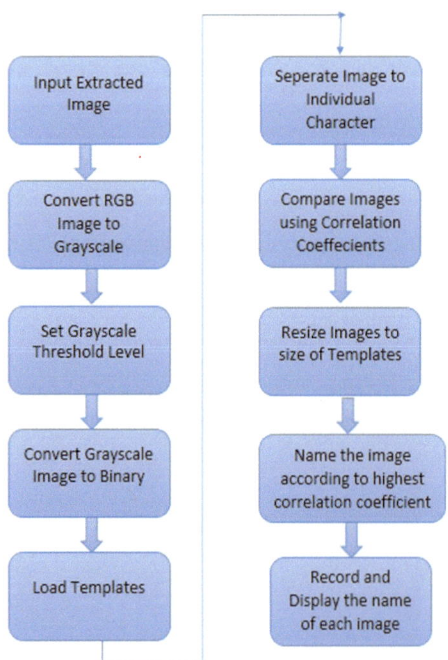

Fig. 3 Predefined templates

remove unwanted noise (anything below 70 pixels). We have used the following two functions for preprocessing:

$$\mathrm{bwareaaopen}(\mathrm{image1}, 70) \quad (1)$$

$$\mathrm{medfitl2}(\mathrm{image1}) \quad (2)$$

We have used edge detection function to acquire number plate region. Next, our algorithm identifies the individual character by using segmentation technique using the following functions:

$$[\mathrm{r}, \mathrm{c}] = \mathrm{find}(\mathrm{image})\%\mathrm{r} \rightarrow \mathrm{rows}, \mathrm{c} \rightarrow \mathrm{column} \quad (3)$$

$$\mathrm{maxc} = \max(\mathrm{c}); \mathrm{minc} = \min(\mathrm{c}) \quad (4)$$

$$\mathrm{maxr} = \max(\mathrm{r}); \mathrm{minr} = \min(\mathrm{r}) \quad (5)$$

$$\mathrm{image1} = \mathrm{image}(\mathrm{minc}\colon \mathrm{maxc}, \mathrm{minr}\colon \mathrm{maxr}) \quad (6)$$

Now after having the images of individual characters, we correlate these obtained images with predefined templates as shown in Fig. 3. The last step is writing the obtained characters (vehicle registration number) into a text file.

Parking Slot Allocation Algorithm This begins with a camera system initialization followed by the acquisition of image of the parking slot. The acquired image is converted into a binary image as explained above, and we have assumed that the parking area is always white (or light color) with black (or dark colored) boundary.

The most important step in this algorithm is image segmentation, which clearly differentiates parking slots from rest of the area, and to check whether the slot is occupied or vacant. From the acquired image of the parking slot, the parking slots are cropped and each cropped parking slot is assigned a respective variable. For each vacant parking slot, we have calculated the total number of pixels; now, when car enters the parking slot, the number of pixels of parking slot area (white area) decreases which we have compared against the certain threshold level to be sure that only the vehicle has occupied the slot not any foreign object.

Depending upon the angle of the camera, a number of pixels for different parking slots vary from each other. Number of Pixels = Number of Rows X Number of Columns.

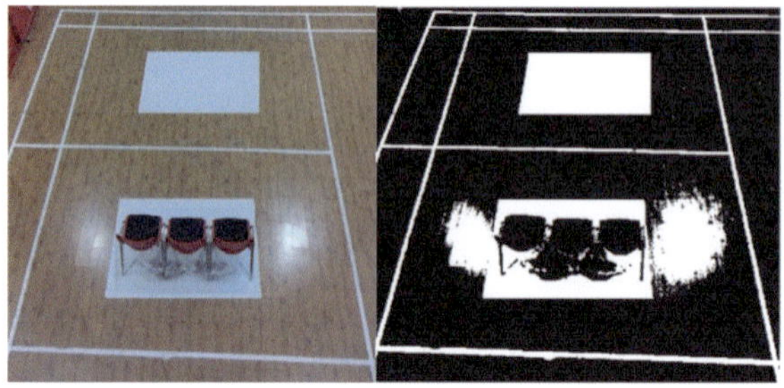

Fig. 4 Original and binary image of pseudoparking slot, chairs emulating a vehicle

We have taken a pseudoparking slot with two parking slots whose images are captured from a certain angle. As a result, the pixels obtained are different for two slots as follows: Slot 1—67,000 and Slot 2—33,000.

Different vehicles have different sizes; also, binary images of same vehicle may differ from one another depending upon the angle of the camera. For our test cases, we have taken a threshold of 28,000 pixels for a vehicle.

Figure 4 is a binary image of our pseudoparking slot. Figure 5 is the flowchart of the program.

Back-End System (Database Management) In back-end system, we need to tie the parking slot number along with the parking floor number with registration number of the vehicle parked on that slot. We are using the concept of file handling in MATLAB. The parking slot is assigned a variable P: It can return two values 0, when vacant, and 1, when occupied. When it returns the value 1, a program is run which assigns the latest occupied parking slot number to latest vehicle that has entered the parking area. File handling is used to open the notepad file which contains the vehicle registration number obtained during Number Plate Recognition.

4.2 Hardware Implementation

The hardware of the project consists of two sets of camera: one in the parking slot and another at the entrance of the parking. A Windows OS computer that processes the obtained images in MATLAB is used. A display unit with GUI which is deployed near the entrance is used to display the availability of vacant slots, and several display units at different floors of building with GUI are installed that will help the users to locate their parked vehicle. All displays are connected to the computer.

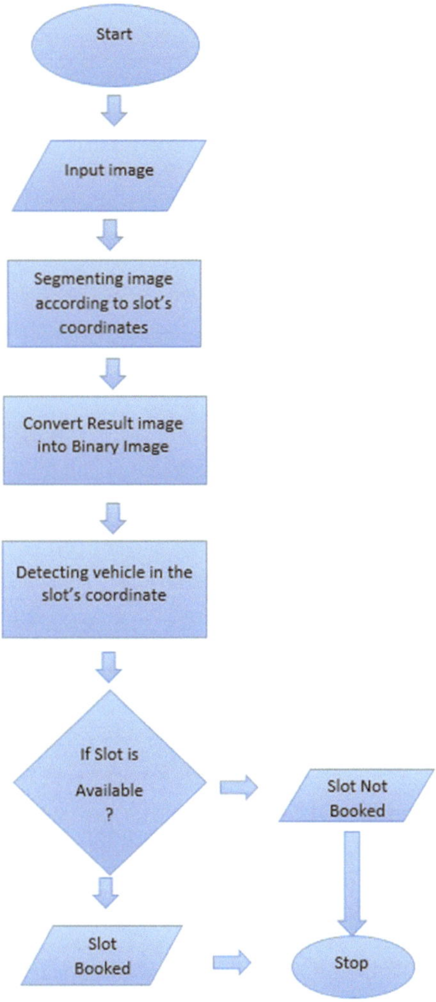

Fig. 5 Flowchart

5 Test Cases and Results

We have considered a pseudounderground parking slot with two parking slots. We have used some objects which act as vehicle. We have a GUI which displays the slot availability. Here are the test cases.

Case 1—Vehicle entering the parking slot and vehicle registration number getting captured

As shown in Figs. 6 and 7, when vehicle enters the parking slot, an image of number plate is captured. Using image processing, the registration number is extracted and stored in the log file with date and time stamp. For each vehicle entering, log file is updated with a new entry.

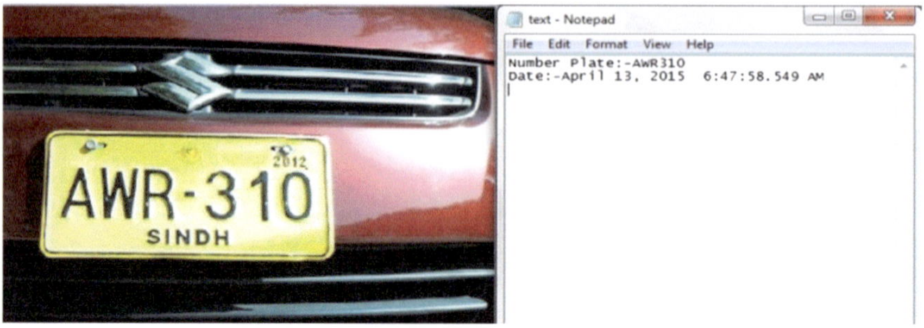

Fig. 6 First car entering the parking slot

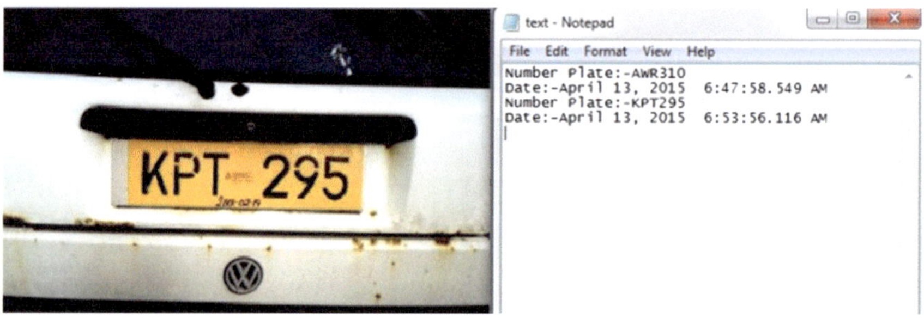

Fig. 7 Second car entering the parking slot

Case 2—GUI displaying the slot status

At the entrance of the parking slot, the real-time information about the nearest parking slot availability is presented on a GUI display. This will help the driver in locating the nearest available parking slot for parking his vehicle. For our test purpose, we have considered two parking slots in the parking slot. When both slots are vacant, GUI will return a numerical value 0 (Fig. 8) indicating to the driver that both slots are available for parking.

When a slot is occupied, GUI will return a numerical value 1, indicating the occupancy of the slot. In Fig. 9, left slot is occupied; in Fig. 10, right-hand-side slot is occupied.

When both slots are occupied as shown in Fig. 11, the value returned in GUI for both slots is 1. We can certainly make our GUI more interactive by replacing the numerical values of 0 and 1 by either text messages or color codes.

Case 3—Slot allocation and database management

The final task is to append the parking slot number (where the vehicle is parked) to the vehicle registration number. This is a very important step, because this will help the owner of the vehicle to locate his/her vehicle from anywhere in the building (e.g., mall, shopping complex). Each floor of the building will be equipped with a GUI display on which owner will enter his vehicle registration number (or a

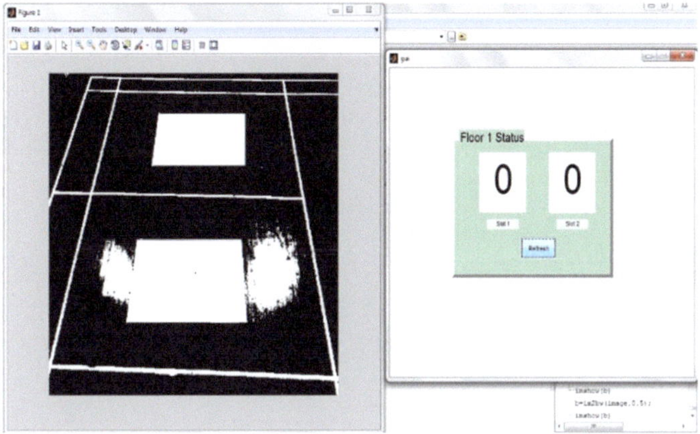

Fig. 8 Both slots vacant, GUI returning 0 for both slots

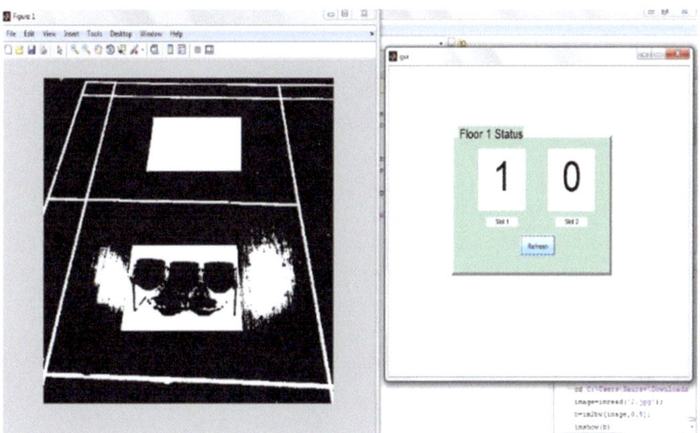

Fig. 9 Slot 1 occupied

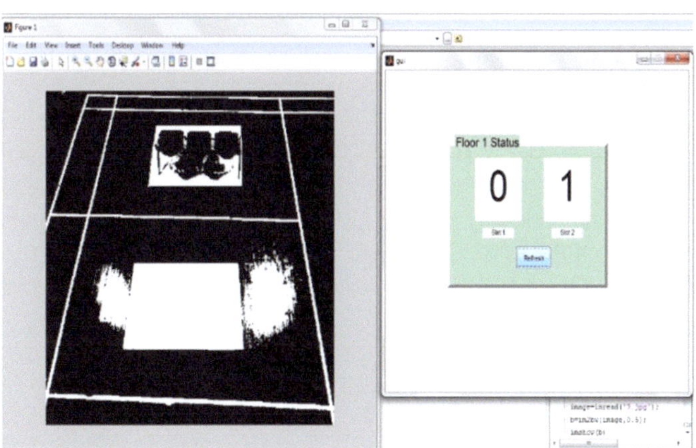

Fig. 10 Slot 2 occupied

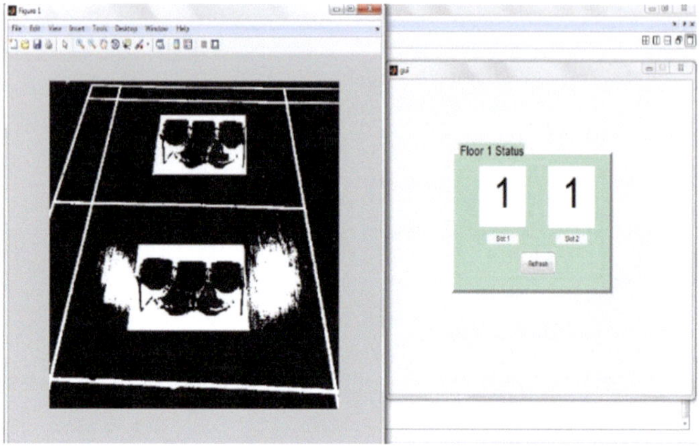

Fig. 11 Both slots occupied

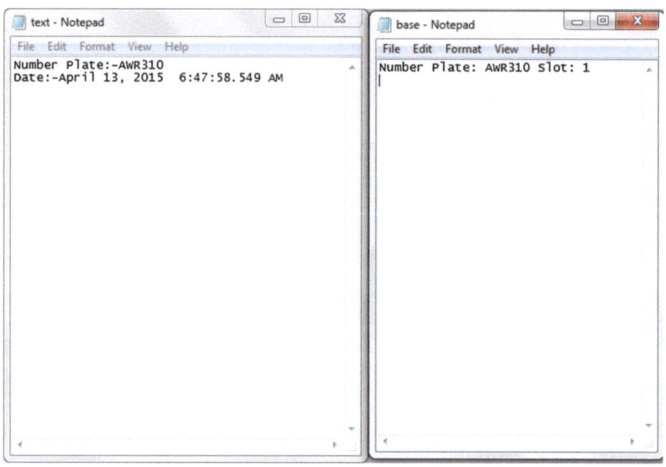

Fig. 12 Slot 1 assigned to the first vehicle

part of registration number) and the system will display the exact parking slot number (location). Figures 12 and 13 show the database where the vehicle registration number is appended to parking slot number and the log file is maintained. Figure 12 shows that the first vehicle with registration number AWR310 is assigned Slot 1, and Fig. 13 shows that the second vehicle with registration number KPT295 is assigned Slot 2.

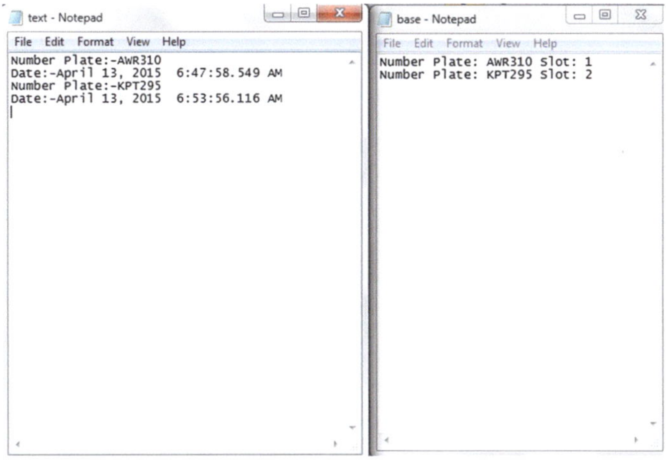

Fig. 13 Slot 2 assigned to the next vehicle

6 Future Scope and Conclusions

6.1 Future Scope

We have tested our system on pseudoparking slot. We plan to either use a bigger parking slot or actual parking slot. Currently, our Number Plate Recognition is working on certain set of images (number plate font, etc.) and we plan to make it universal. In our present work, we have used parking slot as white surface with black marking. We plan to make it universal for the surface of any color. We plan to improve our system of assigning slot number to the vehicle registration number. Lastly, we plan to add automatic ticketing to our system.

6.2 Conclusion

The objective of the project was to automate the existing multistoried parking facilities by employing a low-cost system which requires very minimal dependencies on hardware. This ensures that system is more robust and requires very limited maintenance. Our system is boon to those users who frequently use multistoried parking as for them locating a vacant parking slot for their vehicle and locating their parked vehicle upon return are quite simplified. We have successfully tested our system.

References

1. Managing Parking Issues with Automated Parking Solutions: http://www.moneycontrol.com/master_your_money/stocks_news_consumption.php?cat=realestate&autono=1106851.
2. Parking Problems: http://www.thehindu.com/todays-paper/tp-national/tp-kerala/parking-problems/article4177507.ece.
3. Different Types of Parking Spaces and Multiple Level Car Parking: http://www.engineeringcivil.com/different-types-of-parking-spaces-and-multiple-level-car-parking.html.
4. Automated Car Parking System By PARI: http://paricarparking.com/.
5. Mithari, R., Vaze, S., Sanamdikar, S.: Automatic multistoried car parking system. Inter. J. Innovative Technol. Adapt. Manage. (IJITAM). ISSN: **1**(6), 2347–3622 (2014)
6. Kianpisheh, A., Mustaffa, N., Limtrairut, P., Keikhosrokiani, P.: Smart parking system (SPS) architecture using ultrasonic detector. Inter. J. Softw. Eng. Appl. **6**(Number-3) (2012)
7. Al-Kharusi, H., Al-Bahadly, I.: Intelligent Parking Management System Based on Image Processing. World J. Eng. Technol. **2**(2014), 55–67 (2014)
8. Sutar, G., Shah, A.: Number Plate Recognition using an Improved Segmentation. Inter. J. Innovative Res. Sci. Eng. Technol. ISSN **3**(5), 2319–8753 (2014)

Computer-Aided Detection of Breast Cancer Using Pseudo Zernike Moment as Texture Descriptors

Shabana Urooj, Satya P. Singh and A.Q. Ansari

Abstract The breast cancer is a prominent cause of decease in women worldwide. The early detection of breast cancer may avoid the causing symptoms to spread beyond the breast which can significantly reduce the decease rates. In this paper, we develop a computer-aided diagnosis (CAD) system to detect and classify the abnormalities. The input region of interest (ROI) is manually extracted and subjected to further several preprocessing steps. The pseudo zernike moment (PZM) is used for feature extraction as a texture descriptors. A support vector machine is implemented to classify the extracted features accordingly. The proposed system accomplished overall accuracy of 93.63% with 92.14% sensitivity and 94.14% specificity. The area under the curve (AUC) is found to be 0.974.

Keywords Mammogram · CAD · Pseudo zernike moment · Support vector machine · Segmentation

1 Introduction

Breast cancer is a prominent cause of decease in women worldwide. Breast cancer usually begins in the inner-lining of the milk ducts or the lobules that supply them with milk [1]. This results in malignant cells and the possibility of growing a malignant tumor in another part of the body if not discriminated in its early stage. In the recent years, varieties of computer-aided detection (CAD) systems have been

S. Urooj · S.P. Singh (✉)
Department of Electrical Engineering, School of Engineering,
Gautam Buddha University, Greater Noida 201310, India
e-mail: s.p.singh@ieee.org

S. Urooj
e-mail: shabanaurooj@ieee.org

A.Q. Ansari
Electrical Engineering Department, Jamia Millia Islamia (A Central University),
New Delhi 110025, India

S. Urooj and J. Virmani (eds.), *Sensors and Image Processing*, Advances in Intelligent Systems and Computing 651, https://doi.org/10.1007/978-981-10-6614-6_9

developed to assist the radiologist in automatic differentiation between benign and malignant tumor [2, 12–14]. A combination of CAD and experience of an expert radiologist can significantly improve the accuracy of diagnosis process. In medical imaging, CAD can be classified in two ways: (a) the CAD system points out the region of suspicious (ROS) and alerts the expert radiologist to the need of further analysis and (b) the CAD system which can take the decision from ROS whether it is normal, benign, or malignant tissue. However, the expert radiologist uses CAD system as a second opinion, and the radiologist makes the final decision [3]. The current study is directed toward the development of Computer-aided diagnostic (CAD) system for automatic detection of breast cancer.

2 Motivation

In the mammographic medical X-ray imaging, the expert radiologist visually searches the mammogram for some specific abnormalities. The expert radiologist pays attention to the shape, margin, and density features of the mammograms. However, the human involvement causes a low degree of precision which often results in biopsy and anxiety for the patient involves. Recently, a variety of CAD systems has been developed to assist the radiologist and to reduce the human factor involvement [4–6, 15, 16]. A combination of computer-assisted diagnosis and expertize of radiologist significantly reduce the false negative rates. Recently, Saki et al. [7] propose Zernike moments (ZM) as feature descriptors for mammographic images. However, as a matter of fact, the polynomial of ZM involves many factorial terms and often results in high computational time. In this paper, we use Pseudo Zernike Moment (PZM) as a feature descriptor of mammographic abnormality [8]. PZM have good property of orthogonality and rotation invariance. Furthermore, it has been validated that PZM has better feature representation capability than ZM.

3 Database

In this paper, proposed CAD system is tested and validates on the mammographic image analysis society (MIAS) database [9]. The original MIAS database has been required to 200-micron pixel edge and clipped/padded so that each mammogram size is 1024×1024 pixel. The MIAS database contains 330 digital mammograms belonging to the left and right breast of different women. The x-y coordinates and radius of the abnormalities are already annotated on the mammographic images in the database. In the current study, the region of interest (ROI) is manually segmented using the annotated information and with the help of three experienced radiologists (Fig. 1).

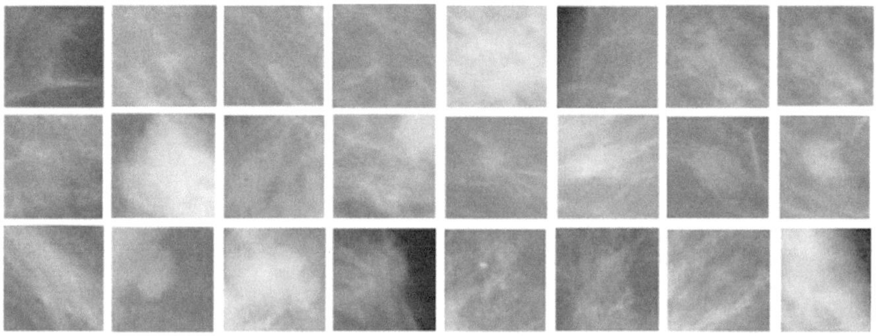

Fig. 1 Examples of extracted ROI. *Top row* from *left to right* (Normal), *middle row* Benign and *bottom row* (Malignant)

4 Preprocessing

The aim of preprocessing is to remove futile and irrelevant information such as non-breast region, background information, pectoral muscles, and several types of noises. The best way to get rid of these artifacts is extracting the ROI manually. Although, this method is time-consuming, it gives quite promising results. In this paper, we extract the ROI of size 128 × 128 pixels manually under the guidance of three experienced radiologists and considering the annotated information of the mammographic image itself. Furthermore, the ROI is enhanced by linear polynomial filtering and subjected to the segmentation using two-level set method [10]. Finally, a circle of 64 × 64 pixels in radius is cropped from the segmented region. Now, the database is ready for feature extraction for specific abnormalities.

5 Feature Extraction and Selection

In this stage, the properties of PZM are utilized for feature extraction. The PZM is used as a texture descriptor of the manually segmented preprocessed suspicious region. The 2-D PZM is a set of complex polynomial which represents a complex orthogonal set over the interior of unit circle defined by $x^2 + y^2 = 1$. Let this polynomial is denoted by $V_{p,q}$, the PZM, $Z_{p,q}$, of order p, and repetition q in polar coordinate is given as:

$$Z_{p,q} = \frac{p+1}{\pi} \int_0^{2\pi} \int_0^1 V^*_{p,q}(r,\theta) f(r,\theta) r \mathrm{d}r \mathrm{d}\theta, \quad p \geqslant 0, \ |q| \leqslant p \qquad (1)$$

where ‘*’ denotes the complex conjugate, ‘r’ is the length of the vector from the origin to a particular pixel (x, y) within the unit circle. The basis function $V_{p,q}$ can be decomposed into the radial and circular components, such as

$$V_{p,q}(r,\theta) = R_{p,q}(r)\exp(jq\theta) \tag{2}$$

where $R_{p,q}$ is the radial polynomial and is defined as:

$$R_{p,q}(r) = \sum \frac{(-1)^k (2p+1-s)!}{s!(p-|q|-s)!(p+|q|+1-s)!} r^{p-s} \tag{3}$$

The basis function of PZM also satisfies the orthogonal property such as

$$\int_0^{2\pi}\int_0^1 V_{p,q}(r,\theta)V^*_{\ell,k}(r,\theta)r\mathrm{d}r\mathrm{d}\theta = \frac{\pi}{p+1}\delta_{p\ell}\delta_{pk} \tag{4}$$

where δ_{nm} denotes the Kronecker delta function.

Now, let the input image is rotated through an angle of α. The PZM of the original image and its rotated version can be related as:

$$Z^{rotated}_{p,q} = \frac{p+1}{\pi}\int_0^{2\pi}\int_0^1 R_{pq}\exp(jq\theta)f(r,\theta-\alpha)r\mathrm{d}r\mathrm{d}\theta \tag{5}$$

By changing the variable $\theta' = \theta - \alpha$

$$Z^{rotated}_{p,q} = \frac{p+1}{\pi}\int_0^{2\pi}\int_0^1 R_{pq}\exp(jq(\theta'+\alpha)f(r,\theta')r\mathrm{d}r\mathrm{d}\theta' \tag{6}$$

$$Z^{rotated}_{p,q} = Z_{pq}\exp(jq\alpha) \tag{7}$$

Equation (7) shows the simple rotational property of PZM. If somehow, we suppress the exponential term of (7), we get completely rotational invariant system. The simplest way to get the absolutely invariant is to take the modulo both side of (7) which results in:

$$\left|Z^{rotated}_{p,q}\right| = \left|Z_{pq}\right| \tag{8}$$

where $Z^{rotated}_{p,q}$ and Z_{pq} are the PZM of the rotated mammogram and the original mammograms, respectively. Thus, our proposed features are the magnitude of the PZM as texture descriptors. These, extracted features are classified into malignant and non-malignant using the appropriate classification scheme as described in the next section.

6 Classification

We use support vector machines (SVM) algorithm to classify the extracted features accordingly. The SVM have been reported in the literature effectively in computer-assisted diagnosis systems for breast cancers. The SVM are based on the supervised learning with associated learning algorithms. The idea behind the SVM is to find an optimal hyperplane in a vector space where the original patterns are embedded via a nonlinear mapping. The details of SVM are omitted here for brevity. The interested readers are referred to the earlier articles [11] for more detail.

7 Results and Discussion

The proposed system is tested and validated in MATLAB R2009b environment on the Windows 7 platform. The suspicious region is manually extracted and passed to several preprocessing stages. The features are extracted from using the proposed PZM. These extracted features are classified into benign and malignant tumor accordingly using SVM as the classifier [11]. The performance of the proposed CAD system is evaluated by considering the number of false positive (FP), false negative (FN), true positive (TP), and true negative (TN).Receiver operating characteristics (ROC) are constructed by plotting true positive rate (TPR) and false positive rate (FPR). Breast Imaging Reporting and Data system (BI-RADS) scoring system is used to set the threshold. The terms, accuracy, sensitivity, and specificity are defined as [2]

$$Accuracy = \frac{TP + TN}{TP + FN + TP + FP} \tag{9}$$

$$Sensitivity = \frac{TP}{TP + FN} \tag{10}$$

$$Speicificity = \frac{TN}{FP + TN} \tag{11}$$

The results are reported in Table 1. The proposed CAD system is compared with the well-known orthogonal moment i.e., ZM-based feature descriptors.

It has been noticed that the proposed system performs well than the ZM-based CAD system in terms of overall accuracy, sensitivity, and accuracy. The overall accuracy of the proposed CAD system is 93.63% with 92.14% sensitivity and 94.14% specificity. On the other hand, the ZM as a feature descriptors on the same database and classification scheme gives the overall accuracy of 89.27% with sensitivity 90.13% and 88.12% specificity. In conclusion, the proposed system performs better the ZM-based feature descriptors in terms of accuracy, sensitivity, and specificity. We also plotted the ROC curve for the proposed system. The results

Table 1 Comparison table of the proposed CAD system with other CAD system

Method	Database	Feature descriptors	Classification	Sensitivit %	Specificit %	Accuracy %
Proposed	MIAS	PZM as a texture descriptors	SVM	92.14	94.14	93.63
Saki et al. [7]	MIAS	ZM AGL, Con, SpI, NRL derivatives, Zernike moments	OWPE	90.13	88.12	89.27

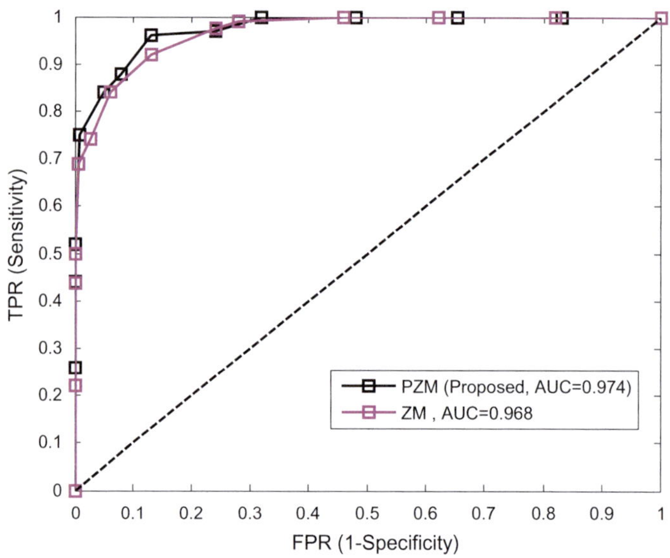

Fig. 2 Comparision analysis of receiver operating characteristics curve of the proposed CAD system

are shown in Fig. 2. The best area under the curve (AUC) for the proposed system is found to be 0.974, and the best AUC for the ZM-based system is found to be 0.968. It is concluded that the proposed feature descriptors for the CAD system perform reasonably well than ZM-based feature descriptors.

8 Conclusion

The work proposes a CAD system to detect and classify the abnormalities in breast cancer. The suspicions region is extracted manually to avoid the processing of whole mammogram. The magnitude of PZM is utilized for feature extraction as texture descriptors. A support vector machine is used for classification purpose. It is concluded that the proposed CAD system performs reasonably well than ZM-based CAD system in terms of accuracy, sensitivity, and specificity. Moreover, the proposed CAD system improves the accuracy of the diagnosis system and promises a role as a second eye of expert radiologist.

References

1. Kelsey, Jennifer L., Marilie D. Gammon, and Esther M. John.: "Reproductive factors and breast cancer." Epidemiologic reviews 15.1 (1993)
2. Urooj, S., and Singh, S.P.: Rotation Invariant Detection of Benign and Malignant Masses Using PHT. IEEE 2nd International Conference on Computing for Sustainable Global Development (INDIACom), 11–13 March 2015, pp. 1627—1632
3. Rouhi, R., Jafari, M., Kasaei, S., Keshavarzian, P.: Benign and malignant breast tumors classification based on region growing and CNN segmentation. Expert Syst. Appl. **42**(3), 990–1002 (2015)
4. Sharma, S., Khanna, P.: Computer-Aided diagnosis of malignant mammograms using zernike moments and SVM. J. Digit. Imaging **28**(1), 77–90 (2015)
5. Dheeba, J., Singh, N.A., Selvi, S.T.: Computer-aided detection of breast cancer on mammograms: A swarm intelligence optimized wavelet neural network approach. J. Biomed. Inform. **49**, 45–52 (2014)
6. Jalalian, A., Mashohor, S.B., Mahmud, H.R., Saripan, M.I.B., Ramli, A.R.B., Karasfi, B.: Computer-aided detection/diagnosis of breast cancer in mammography and ultrasound: a review. Clin. Imaging **37**(3), 420–426 (2013)
7. Saki, F., Tahmasbi, A., Soltanian-Zadeh, H., Shokouhi, S.B.: Fast opposite weight learning rules with application in breast cancer diagnosis. Comput. Biol. Med. **43**(1), 32–41 (2013)
8. Dai, Xiubin, Liu, Tianliang, Shu, Huazhong, Luo, Limin: Pseudo-Zernike moment invariants to blur degradation and their use in image recognition, pp. 90–97. Berlin, Intelligent Science and Intelligent Data Engineering. Springer (2013)
9. Suckling, J., et al.: The Mammographic Image Analysis Society Digital Mammogram Database. ExerptaMedica Int. Congr. Ser. **1994**, 375–378 (1069)
10. Li, C., Huang, R., Ding, Z., Gatenby, J.C., Metaxas, D.N., Gore, J.C.: A level set method for image segmentation in the presence of intensity inhomogeneities with application to MRI. Image Proc. IEEE Trans. **20**(7), 2007–2016 (2011)
11. Sharma, S., Khanna, P.: Computer-Aided diagnosis of malignant mammograms using zernike moments and SVM support vector machine. J. digital imaging **28**, 77–90 (2015)
12. S.P Singh, S. Urooj, "Combined Rotation- and Scale-Invariant Texture Analysis Using Radon-Based Polar Complex Exponential Transform", Arab. J. Sci. Eng. April (2015)
13. Satya, P.: Singh, Shabana Urooj, "Rotational-Invariant Texture Analysis Using Radon and Polar Complex Exponential Transform". Adv. Intell. Syst. Comput. **327**, 325–333 (2015)

14. Satya P Singh, Shabana Urooj, Aime Lay Ekuakille, "Rotational-Invariant Texture Analysis Using Radon and Polar Complex Exponential Transform", FICTA 2014;Series Title Advances in Intelligent Systems & Computing, Springer-International Publishing Switzerland. doi:10.1007/978-3-319-11933-5_35
15. Bhateja, V.: Shabana Urooj, M Mishra, "Technical Advancements to Mobile Mammography using non-linear Polynomial Filters and IEEE 21451 NCAP Information Model". IEEE Sens. J. (2014). doi:10.1109/JSEN.2014.2366599
16. Vikrant B, Mukul M, Shabana Urooj, "A Robust Polynomial Filtering Framework for Mammographic Image Enhancement from Biomedical Sensors" IEEE Sens. J. **13**, 11. doi:10.1109/JSEN.2013.2279003

Analysis of Leukoderma Images Using Neuro-Fuzzy Hybrid Technique

Sudhakar Singh, Shabana Urooj and Satya P. Singh

Abstract This paper presents a novel method to analyze Leukoderma images using Neuro-Fuzzy hybrid (NFH) approach. Skin diseases are the most widespread diseases in India and worldwide. In the proposed work, a hybrid Artificial Neural Fuzzy Inference System (ANFIS) is designed. The advantage of the proposed system is that there is not any connection between fuzzy and neural network. The training data is grouped into several clusters. Each cluster is designed to represent a particular rule. Error rate, output data, and trained data are calculated.

Keywords Leukoderma · Skin disease · Neuro-Fuzzy hybrid

1 Introduction

Skin diseases are one of the foremost causes of human illness worldwide. Skin illnesses like tanning, pigment darkening, sunburn, skin cancers, and infective diseases are growing more rapidly due to litter, ultraviolet light, and global warming. India is a somewhat most polluted country in the world. Pollution increases the risk of Eczema and Psoriasis being the major contributors to the skin disease. Eczema causes the itching on the affected part of the body. It may results in moisture out and germs in other parts of the body if not treated in its early stage. Psoriasis is a kind of skin disease where skin cells move toward the skin surface in days rather than week which results in thick patches called plaques. Psoriasis generally appears on elbow, knee, feet, and scalps or lower back. Adults are mostly

S. Singh (✉) · S. Urooj · S.P. Singh
Department of Electrical Engineering, School of Engineering,
Gautam Buddha University, Greater Noida, India201310
e-mail: sudhakarsingh86@gmail.com

S. Urooj
e-mail: shabanaurooj@ieee.org

S.P. Singh
e-mail: satya002u@gmail.com

S. Urooj and J. Virmani (eds.), *Sensors and Image Processing*, Advances in Intelligent Systems and Computing 651, https://doi.org/10.1007/978-981-10-6614-6_10

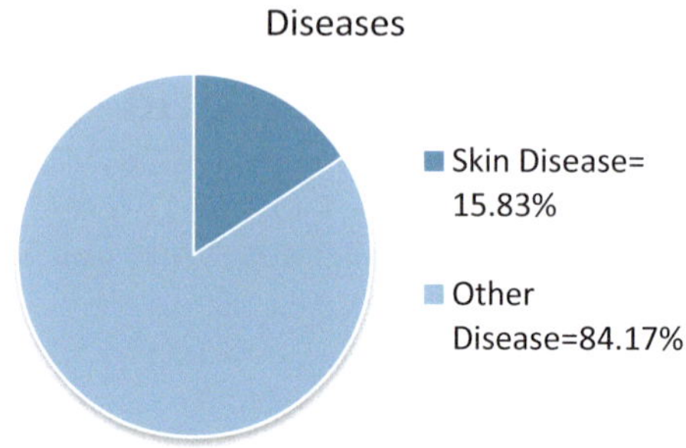

Fig. 1 Distribution of diseases

affected with psoriasis. It is found that psoriasis causes due to overreacting of immune system results in inflammation and flaking of skin.

It is evident that the ozone layer is highly affected by global warming and pollution. One percent decline in ozone leads to a 2–5% boost in the prevalence of skin cancers and other related disease. The natural balance for skin care cure is endorsing an excessive transform in requirement patterns. Short-lived with the days, skin diseases care was tapered to supervision of regular men. The developments of ornamental and antiaging treatments have tainted the face of the skin care industry. However, these treatments are the major cause of skin disease. The number of these patterned measures is expected reach to 18–20 lakhs by 2015 in India. In 2013, it was with occurrence pace of 10 percent, the population affected across India from skin disease is estimated at nearly 15.1 crores. It is expected that at a CAGR of 12 percent with reference to 18.8 crores people are probable to suffer from skin disease by 2015 in India (Fig. 1).

Computer-aided detection (CAD) can be used on the skin disease images to help the skin expert radiologist as a second reader. CAD can find and differentiate the skin disease that a skin disease expert may not spot [17, 18]. Once the CAD analysis has been done, the skin expert can take the decision based on the classification results. In this manuscript, I planned a fully automatic computer-aided system to help the skin expert in the identification of the skin disease such as Leukoderma skin images.

2 Literature Review

Detection of eczema in early stage is the most important part because eczema can be cured by means of a simple cutting out. The revelation of the skin to rays is considered the main cause of the growth of the skin cancers. Dermoscopy is a principally extremely helpful standard technique to diagnosing the spiteful cells of skin disease [1]. The vital signs of dermoscopy are broaden in correctness matched with naked-eye examination (up to 28% in the case of sensitivity and up to 14% in the case of specificity), thereby sinking the occurrence of pointless surgical removals of benign lesions [2, 3]. A number of symptoms may be ignored or overvalued due to the role of human factor which results in without cause cost and treatment to the patient.

There are practically nearly 3000 deaths from skin infections, more than 2000 from malicious melanoma, and nearly 700 from non-malignant malignancy. In 2008, it was manifest in the USA that about 60,000 people were feeling pain with dangerous skin disease and about 9000 people died skin diseases. If I talk about Australia, it was calculated that additional 12,500 fresh personal belongings of skin infections are testified each year, out of which about 2000 cases are cause of decease. Australia has the highest speed of occurrence of pelt illness in the world. It was calculated that about 3000 deaths from skin diseases, more than 2000 from malignancy, and 620 from non-malignant tumor.

Malicious growth is mainly caused by means of UVA radiation passing through the indirect DNA harmed. Skin infection is a large amount added unsafe if it is not found in the premature stages. It causes the mainstream (77%) of deaths from the skin cancer [4]. Internationally, in 2012, malignancy cases identified were about 250,000 people and consequence in 55,000 deaths.

In the past two decades, many researchers have been concerned with improving fuzzy logic techniques used for color investigation as well as segmentation in medicinal applications. A number of the fuzzy logic techniques examined inside preceding research used for color and furthermore skin assessment including color histogram examination for color marking and skin laceration intolerance [5], defect range assessment for fur abrasion unfairness [6], constraint recognition are also defined in skin lesion using fuzzy clustering and texture analysis [7]; fuzzy clustering segmentation technique is used for adaptive removing background of skin color for part segmentation [8], fuzzy c-means clustering considered for pelt scratch analysis [9], and adaptive fuzzy c-means is improved technique using local spatial continuity for cluster pattern evaluation [10], here skin area segmentation by fuzzy estimation grade modeling [11]. The hypothesis of fuzzy logic sets is an entrancing as well as a beneficial tool, while it avails a high-quality theoretic commencement in the direction of exemplifying fuzziness in sequence in addition to it constitutes an incorporated structure for instead of and giving out both numerical and symbolic objects, in addition to systematic information. Scheduled starting the foundation of artificial intelligence, soft computing, and image processing, used for design as of [12–16], a number of authors have second-hand these techniques to give support to investigation of medical images; novel segmentation method be single of the distinctive technique because it has to be desired a model image of the analyzed configuration. Furthermore, the addition of area understanding concerning the individuality of objects increases the segmentation precision [17, 18, 19]. The shape of fuzzy sets considered in favor of this purpose makes use of a rule-based technique narrative of the general idea relating to the Region of Interest that makes accessible to channel, the system for the edge-detection-based segmentation analogous to [20]. On the opening of the scheme, used for every input image, a representative image is strong-minded. After that, fuzzy rules are functional among every single points of thoughtfulness (applicant points used for shape detection), single individuals substantial the rule sets are reserved for investigation, by applying a number of criterion of decimation.

3 Materials and Methods

The methodology of this work is based on the combination of neural network, and the advantages of Fuzzy logic called Neuro-Fuzzy hybrid systems (NFHS). NFHS is a kind of neural networks where fuzzy structure is interpreted as an out of the ordinary kind of neural network. There are two major reasons to use NFH system in proposed work. First, the neural networks are a low-level computational network which performs fine among raw information. Second, the fuzzy system cannot learn itself and also can not adjust in new environment. Therefore, by combining these two algorithms, we create a hybrid network which is more crystal clear and able to learn. The basic structure of the proposed system is shown in Fig. 2. The input layer 1 transmits the external crisp signal to the next layer. The fuzzyfication neuron in layer 2 receives the crisp signals form layer 1 and determines the degree to which they belong. Each neurons in the layer 3 belongs to the single rule, and output membership is represented by layer 4. Finally, defuzzyfication layer is the last fifth layer, which is the single output of the hybrid system.

The Neuro-Fuzzy-Hybrid system is generally treated as multineural network. The NFH system be able to used to extend IF-THEN rules which determines the membership function for the input and output relations. Membership function expressing the linguistic terms of the inference rules should be formulated for building fuzzy controller. There are different techniques to design and modeled Neuro-fuzzy hybrid systems. In this case, training data is grouped into several

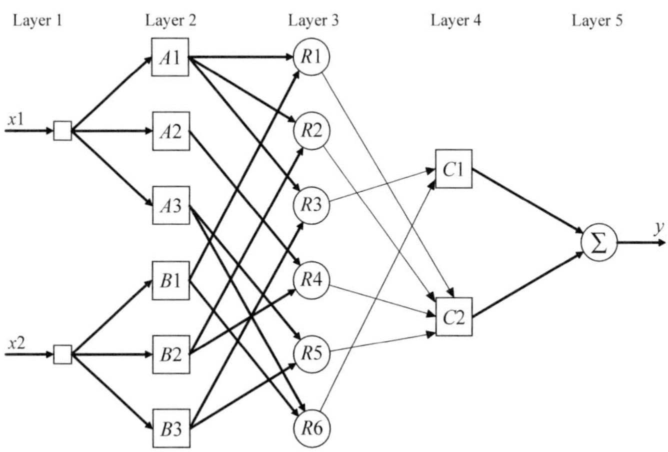

Fig. 2 The basic architecture of the hybrid neuro- fuzzy system

clusters and each cluster is designed to represent a particular rule. The testing can be carried out by presenting a random testing sample to trained neural network. Each and all yield element will go back a level which extends to in shape the precursor of rule. The NFH system uses a learning algorithm consequential inspired by neural network theory to determine its parameters by processing data samples. This system has combined advantages of fuzzy and neural network. Neural network learning provides a good technique to adjust the knowledge of the expert and automatically create supplementary fuzzy rules and membership function to meet certain outputs. Fuzzy sets are considered as weights, whereas the input and output variables in addition to the system are modeled as neurons. Neurons can be incorporated or deleted in the learning phase. Fuzzy knowledge base is represented by neurons. Hence, the main failures of both essential systems are thus climbing beyond. Let the training data $\{(xi,yi) \mid xi \in X, yi \in Y, i = 1,2,\ldots,l\}$ are combined into M clusters. Each cluster represents a rule Rm where m = 1,2,..., M. Every output unit m will go again a degree to which extend x may fit the antecedent of rule Rm. Fuzzy rules are designed on the basis of image parameters such as color value, threshold value, intensity value. Membership functions mf1, mf2, mf3 are decided on the bases of parameters of the image. All the rules of Neuro-fuzzy are designed by parameters of image.

4 Experimental Results and Discussion

Projected algorithm is evaluated on a set of Leukoderma images obtained from http://dermnetnz.org. The sample test image is shown in Fig. 3. The data-set of images of 8-bit color images in JPEG format with 110 × 164-pixel size. I tested more than 100 test images. The gray scale edition of the test image is shown in Fig. 4.

The magnitude of the gradient stretches the facts how quickly the image is changing, while the direction of the gradient expresses us the track in which the image is varying most rapidly (Figs. 5 and 6). The gradient has a direction as well as magnitude. The information gradient direction and magnitude are converted into feature vector. The gradient may be different at every location. We represent it with a different vector at every image location. The image gradient is a key factor in

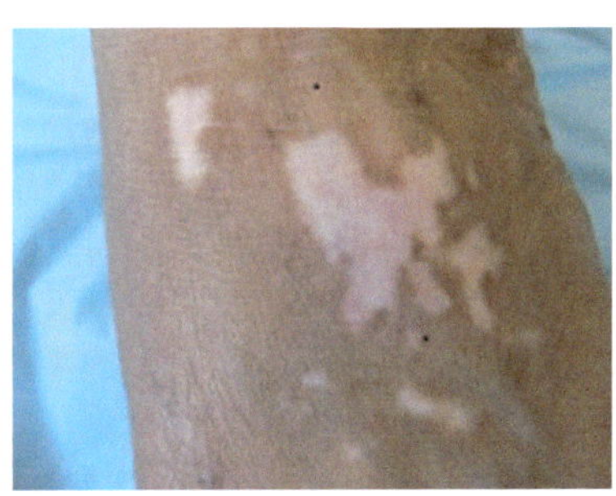

Fig. 3 Show the test image, gray scale image, gradient magnitude, and gradient direction of image using Sobel method and directional gradient Gx and Gy of image using Sobel method, respectively

Fig. 4 Show the test image, gray scale image, gradient magnitude, and gradient direction of image using Sobel method and directional gradient Gx and Gy of image using Sobel method, respectively

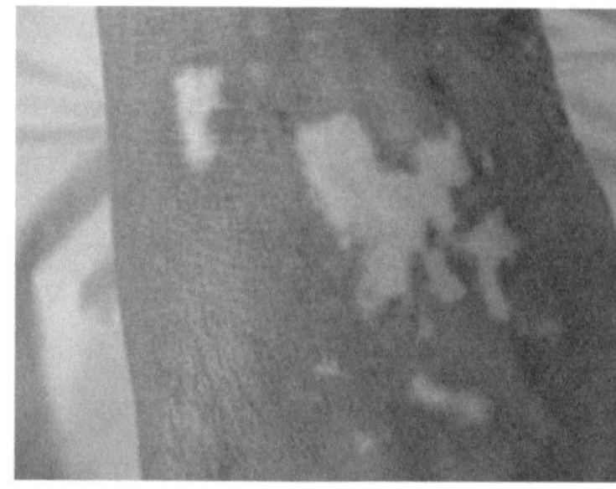

Fig. 5 Show the test image, gray scale image, gradient magnitude, and gradient direction of image using Sobel method and directional gradient Gx and Gy of image using Sobel method, respectively

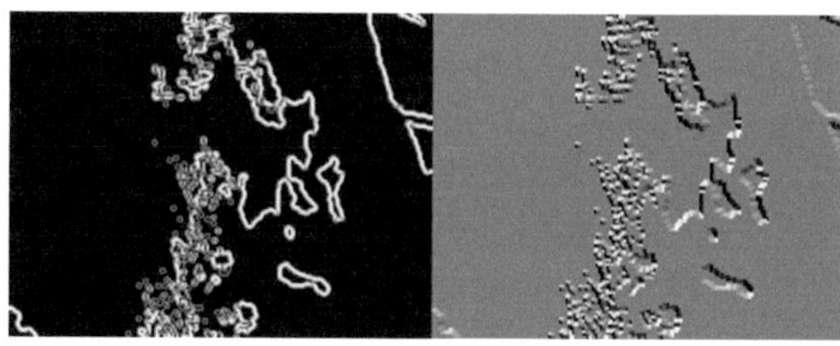

boundary detection because images often change most quickly at the boundary between objects.

Figure 5 shows the segmented image of the test image by Sobel method. In this diagram, gradient magnitude and gradient direction are determined. The direction components along the x-axis and y-axis are shown in Fig. 6. Figure 7 shows the surface view of anfis. Figure 8 shows fuzzy rules which are applying for the analysis. Figure 9 shows the analysis error. The analytic error is calculated by the difference between target and evaluated anfis. From Fig. 9, it could be observed that the error decreases as we move away. The error rate varied very speedily, but after some time with respect to time, the error started to get slow and reach near to zero. Figure 10 shows the output vs trained data. From the fig, we can see the trained data and output data coincide with each other at the same point. It means the accuracy of the proposed technique is very good, obtained result from the proposed technique is very accurate, and error is very less when observation is taken.

Fig. 6 show the test image, gray scale image, gradient magnitude, and gradient direction of image using Sobel method and directional gradient Gx and Gy of image using Sobel method, respectively

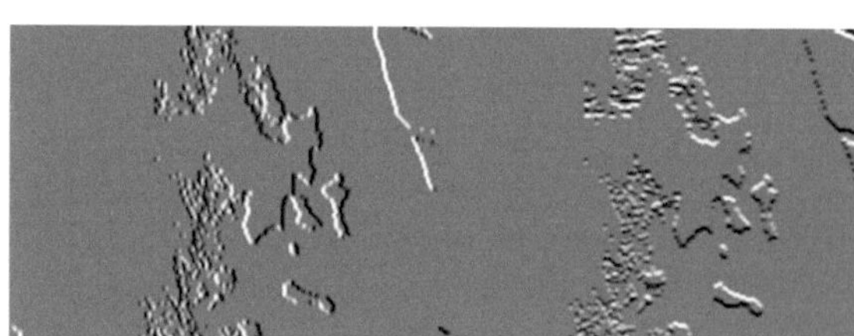

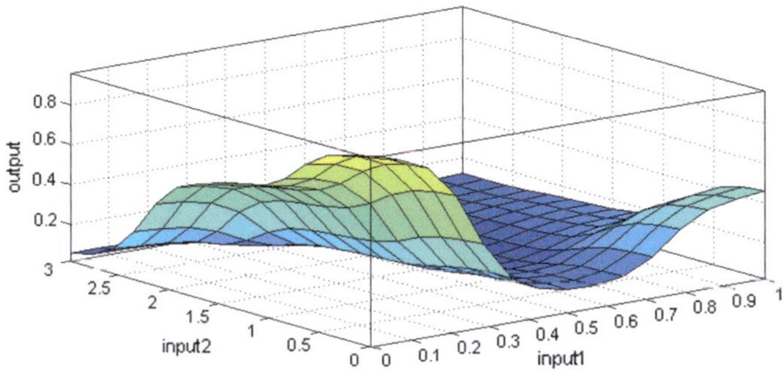

Fig. 7 Surface view of anfis

Fig. 8 Fuizzy rule

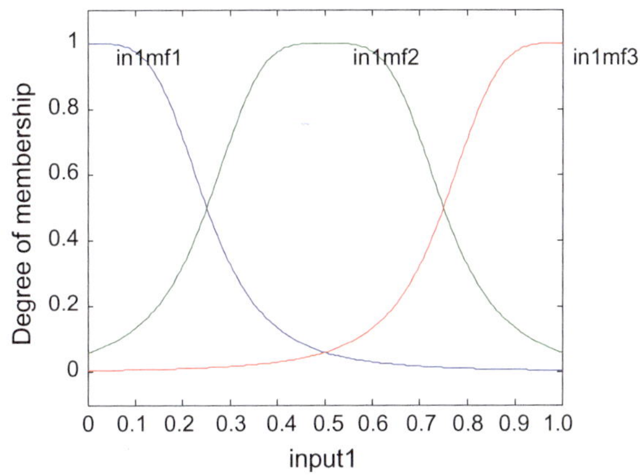

Fig. 9 Error rate

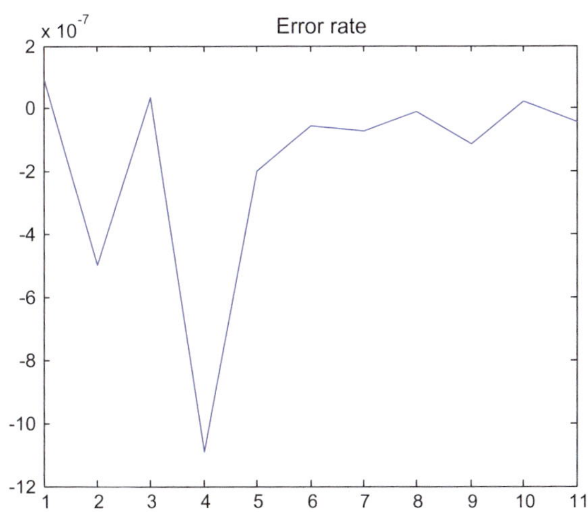

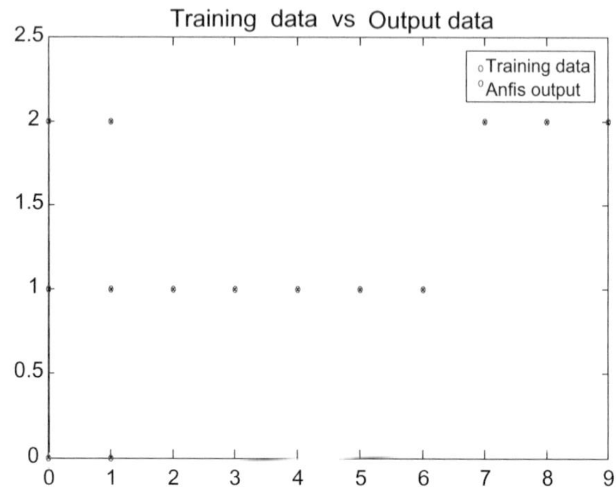

Fig. 10 Train data vs output data

5 Conclusion

In this work, we analyze the Leukoderma images using the proposed Hybrid Neuro-fuzzy system. The proposed technique we used in Leukoderma images. Figure 6 shows gradient magnitude and gradient direction, Fig. 7 shows gradient directional using Sobel method, Fig. 8 shows the degree of membership of the Leukoderma images, Fig. 9 shows error rate, Fig. 10 describes train data and output data. Proposed methodology of computer-aided for Leukoderma maintain merges medical understanding with a number of cutting-edge technologies: image processing, pattern classification, statistical learning, and ensemble techniques of model-based classifiers.

References

1. Argenziano, G., Soyer, H.P.: Dermoscopy of pigmented skin lesions—a valuable tool for early diagnosis of melanoma. Lancet Oncology **2**(7), 443–449 (2001)
2. Vestergaard, M.E., Macaskill, P., Holt, P.E., Menzies, S.W.: Dermoscopy compared with naked eye examination for the diagnosis of primary melanoma: A meta-analysis of studies performed in a clinical setting. Br. J. Dermatol. **159**(3), 669–676 (2008)
3. Ascierto, P.A., Palmieri, G., Celentano, E., et al.: Sensitivity and specificity of epiluminescence microscopy: evaluation on a sample of 2731 excised cutaneous pigmented lesions. Br. J. Dermatol. **142**(5), 893–898 (2000)
4. World Cancer Report, World Health Organization: 2014. pp. Chapter 5.14. ISBN 9283204298, (2014)
5. Stanley, R.J., Moss, R.H., Van Stoecker, W., Aggarwal, C.: A fuzzy-based histogram analysis technique for skin lesion discrimination in dermatology clinical images. Comput. Med. Imaging Graph. **27**(5), 387–396 (2003)

6. Khan, A., Gupta, K., Stanley, R.J., Stoecker, W.V., Moss, R.H., Argenziano, G., et al.: Fuzzy logic techniques for blotch feature evaluation in dermoscopy images. Comput. Med. Imaging Graph. **33**(1), 50–57 (2009)
7. Madasu VK, Lowell BC. Blotch detection in pigmented skin lesions using fuzzy co-clustering and texture segmentation. Proc. Conf. Digi. Image Comput. Techn. Appl. (DICTA'09). pp. 25–31 (2009)
8. Xiao K, Danghu L, Lansun S. Segmentation of skin color regions based on fuzzy cluster. In: Proceedings of the Symposium on Intelligent Multimedia, Video and Speech Processing. pp. 125–8 (2004)
9. Schmid, P.: Segmentation of digitized dermatoscopic images by two-dimensional color clustering. IEEE Trans. Med. Image **18**(2), 164–171 (1999)
10. Liew AW-C, Yan H, Law NF. Image segmentation based on adaptive cluster proto type estimation. IEEE Trans. Fuzzy Sys. **13**(4), 444–53 (2005)
11. Bhatt RB, Sharma G, Dhall A, Chaudhury S. Efficient skin region segmentation using low complexity fuzzy decision tree model. Proc. IEEE India Conf. (INDICON). pp. 1–4 (2009)
12. Bezdek, J.C., et al.: Fuzzy Models and Algorithms for Pattern Recognition and Image Processing (The Handbook of Fuzzy Sets). Springer, Berlin (2005)
13. Etienne, E.K., Nachtegael, M. (eds.): Fuzzy techniques in image processing. Physica-Verlag, N.Y. (2000)
14. Gonzales, R.C., Woods, R.E.: Digital image processing, 2nd edn. Prentice Hall, New Jersey (2001)
15. Rangayyan, R.M.: Biomedical image analysis. CRC Press, Boca Raton (2005)
16. Semmlow, J.L.; Biosignal and Biomedical Image Processing MATLAB-Based Applications, M. Dekker, (2004)
17. Urooj, S., & Singh, S.P.: Rotation invariant detection of benign and malignant masses using PHT. IEEE 2nd Int. Conf. Comput. Sustain. Global Dev. (INDIACom), 11–13 March, pp. 1627—1632 (2015)
18. Satya, P.: Singh, Shabana Urooj, "Rotational-Invariant Texture Analysis Using Radon and Polar Complex Exponential Transform". Adv. Intell. Sys. Comput. **327**, 325–333 (2015)
19. Singh, S. P., & Urooj, S.: Combined rotation-and scale-invariant texture analysis using radon-based polar complex exponential transform. Arab. J. Sci. Eng. 1–14 (2015)
20. S. Urooj, S. Singh, "A novel computer assisted approach for diagnosis of skin disease", IEEE Int. Conf. 11–13 March INDIACom (2015)

Texture and Clustering-based Skin Disease Classification

Pradeep Mullangi, Y. Srinivasa Rao and Pushpa Kotipalli

Abstract The main objective of this paper is to classify the skin diseases using image classification methods. For examining the texture of the image, a statistical method gray-level co-occurrence matrix (GLCM) was used. GLCM considers the spatial relationship of pixels and characterizes texture of an image by calculating how often pairs of the pixel with specific values and specified spatial relationship occur in an image. The presented work here is focused on the extraction of GLCM features inclusive of contrast, correlation, homogeneity, and energy. Fuzzy c-means clustering along with GLCM is proposed for reducing the time taken for skin disease classification. With simulation results, it is shown that the proposed method is more efficient than GLCM alone method.

Keywords Image classification · GLCM · Fuzzy c-means clustering · Skin disease

1 Introduction

Skin disorders are among the most common diseases in both developing and industrialized countries. People living with skin disease experience stressful life as skin disease affects their confidence and self-esteem in so many different ways. In

P. Mullangi (✉) · P. Kotipalli
ECE Department, Shri Vishnu Engineering College for Women (A),
Andhra Pradesh, India
e-mail: pradeepm999@gmail.com

P. Kotipalli
e-mail: pushpak@svecw.edu.in

Y. Srinivasa Rao
Instrument Technology Department, A.U. College of Engineering (A),
Andhra Pradesh, India
e-mail: srinniwasarau@gmail.com

S. Urooj and J. Virmani (eds.), *Sensors and Image Processing*, Advances in Intelligent Systems and Computing 651, https://doi.org/10.1007/978-981-10-6614-6_11

2013, with a prevalence rate of 10%, the population affected across India from skin disease is estimated at nearly 15.1 crores. The situation is further worsened by the low availability of dermatologists in India [1]. In this scenario, medical imaging plays an important role in quick decision making in skin disease identification. Medical imaging is used for revealing internal structures hidden by the skin as well as to diagnose and treat disease. This paper discusses detection of skin diseases using texture and clustering-based image classification.

Sparavigna [2] proposed a texture-based method in which differences in color and coarseness of skin are quantitatively evaluated by using a statistical approach to the pattern recognition. Abbadi et al. [3] developed a skin disease recognition method based on GLCM and skin color. In this paper, GLCM is used for measuring texture features such as correlation, contrast, energy, and homogeneity. Then, Fuzzy c-means clustering technique is used with texture features extracted by GLCM.

The remainder of this paper is organized as follows. Proposed method is presented in Sect. 2. Extraction of image features using GLCM is discussed in Sect. 3. Fuzzy c-means clustering is introduced in Sect. 4. Section 5 presents simulation results and Sect. 6 concludes the paper.

2 Proposed Method

This method is developed for classification of skin diseases by analyzing textures obtained from a collection of images using features based on GLCM. This method has two phases: (a) Training phase and (b) Classification phase. The methodology of proposed method is shown in the Flow chart (Fig. 1).

We considered extraction of four texture features of an image: Correlation, Contrast, Homogeneity, and Energy. In training phase, the texture features are extracted from a given set of training images. In classification phase, the given test image is segmented and then, the above-mentioned texture features are extracted. Then, Fuzzy c-means clustering is applied on the texture features obtained from training images. Then, Chi. square distances are calculated for skin disease classification.

Computing GLCM and extracting texture features of the image are discussed in the section that follows.

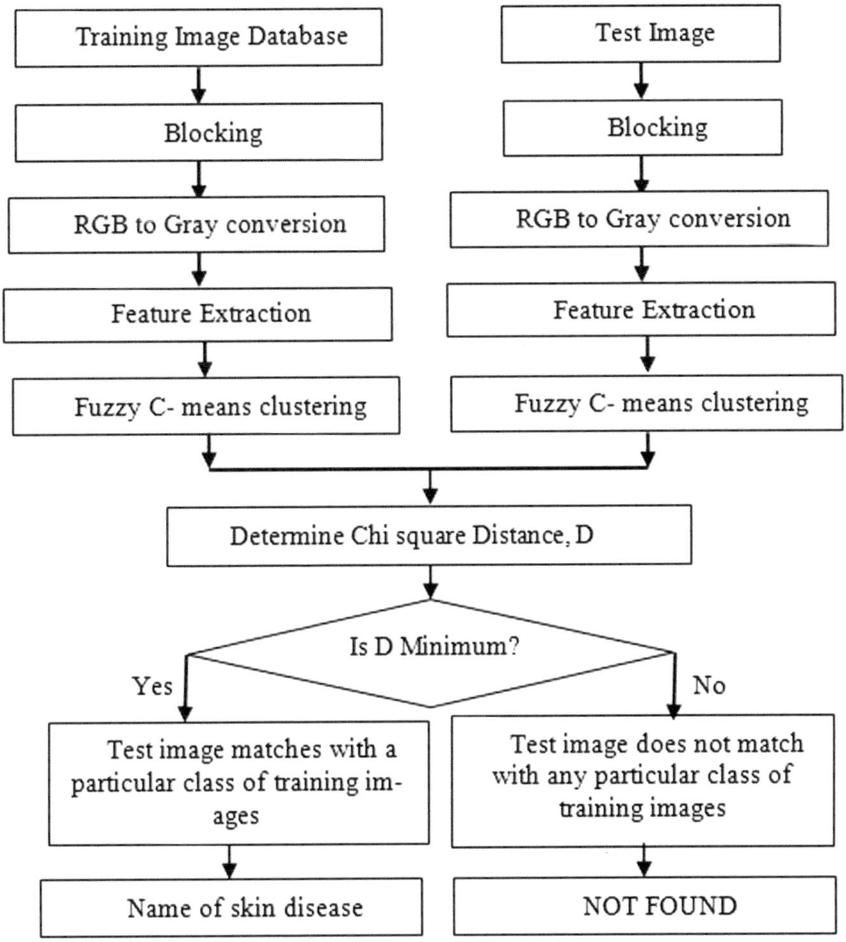

Fig. 1 Flow chart of proposed method

3 Gray-Level Co-occurrence Matrix

GLCM contains the second-order statistical information of spatial relationship of pixels of an image. Texture feature calculations use the contents of the GLCM to give a measure of the variation in intensity at the pixel of interest [4]. In computing GLCM of an image two parameters, *Offset* and *Distance* d between pixels are considered. Here, offset represents the direction of pairing pixels. For example, with 0° offset and $d = 1$ represent pairing horizontally adjacent elements. In this paper, $d = 1$ and four offsets 0°, 45°, 90°, and 135° are considered. Offsets 45°, 90°, and 135° are obtained by pairing the right diagonal elements, vertically upward elements and left diagonal elements, respectively.

3.1 GLCM of an Image

Let us consider a 4 × 4 section of an image having 4 gray levels as shown in Fig. 2a. Image data and GLCM with 0° offset are shown in Fig. 2b, 2c, respectively.

Let G_{0°, G_{45°, G_{90°, *and* G_{135° represent GLCMs with offsets 0°, 45°, 90°, and 135°, respectively. To get the symmetrical GLCM, original GLCM is added to its transpose. Then, normalized symmetrical GLCM is obtained by dividing it with sum its elements. Normalized symmetrical GLCMs N_{0°, N_{45°, N_{90°, *and* N_{135°, respectively are computed.

$$N_{0^\circ} = \frac{1}{22}\begin{bmatrix} 2 & 0 & 0 & 1 \\ 0 & 4 & 3 & 0 \\ 0 & 3 & 4 & 0 \\ 1 & 0 & 0 & 4 \end{bmatrix} \quad N_{45^\circ} = \frac{1}{18}\begin{bmatrix} 0 & 2 & 1 & 0 \\ 2 & 2 & 1 & 2 \\ 1 & 1 & 2 & 0 \\ 0 & 2 & 0 & 2 \end{bmatrix}$$

$$N_{90^\circ} = \frac{1}{24}\begin{bmatrix} 0 & 3 & 1 & 0 \\ 3 & 2 & 2 & 2 \\ 1 & 2 & 2 & 0 \\ 0 & 2 & 0 & 4 \end{bmatrix} \quad N_{135^\circ} = \frac{1}{20}\begin{bmatrix} 0 & 2 & 0 & 1 \\ 2 & 0 & 2 & 1 \\ 0 & 2 & 2 & 2 \\ 1 & 1 & 2 & 2 \end{bmatrix}$$

Then, texture features like Correlation, Contrast, Energy, and Homogeneity of the given image are extracted from normalized symmetrical GLCMs.

3.2 Texture Measures from GLCM

Most of the texture calculations are weighted averages of the normalized symmetrical GLCM cell contents.

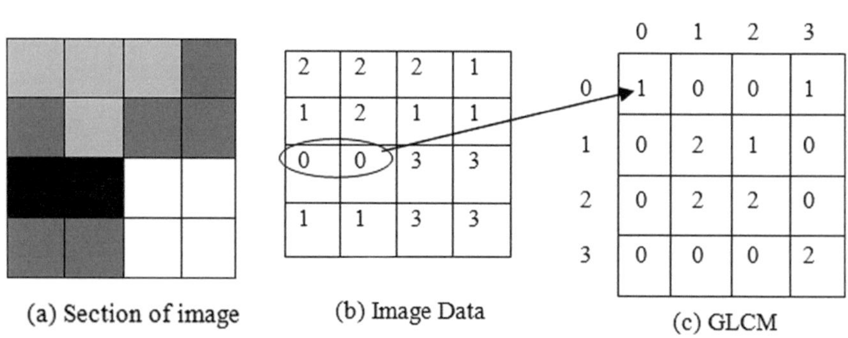

Fig. 2 **a** Small Section of image having 4 gray levels, **b** Image data and **c** GLCM of the image with 0° offset and $d = 1$

Contrast (Con) Contrast is a measure of intensity or gray-level variations between the reference pixel and its neighbor.

$$Con = \sum_{i,j=0}^{K} p_{i,j}(i-j)^2$$

where $p_{i,j}$ represents element (i, j) of the normalized symmetrical GLCM, K represents a number of gray levels in the image and $(i-j)^2$ represents the weight.

Correlation (Cor) It presents how a reference pixel is related to its neighbor. It has values from −1 to 1. −1 refers to uncorrelated, and 1 is perfectly correlated.

$$Cor = \sum_{i,j=0}^{K} \frac{(i-\mu_i)(j-\mu_j)p_{i,j}}{\sigma_i \sigma_j}$$

μ_i, μ_j and σ_i, σ_j represent the means and standard deviations of $p_{i,j}$, respectively.

Homogeneity (*H*) Homogeneity returns a value that measures the closeness of the distribution of elements in the GLCM to the GLCM diagonal.

$$H = \sum_{i,j=1}^{k} \frac{p_{i,j}}{1+|i-j|}$$

Energy (*E*) It measures the textural uniformity that is pixel pair repetitions. It detects disorders in textures. It simply returns the sum of squared elements in the GLCM.

$$E = \sum_{i,j=1}^{K} p_{i,j}^2$$

3.3 GLCM-Based Classification

Let $(N_{0°}, N_{45°}, N_{90°}, N_{135°})_{Tn}$ represent set of normalized symmetrical GLCMs for the training image T_n; $n = 1, 2, \ldots, N$. For each set of normalized symmetrical GLCMs, above-mentioned four texture features are computed and final values of texture features are obtained by taking average of individual feature set of each image [5]. Let $(Cor, Con, H, E)_{Tn}$ represent final values of texture features of training image T_n; $n = 1, 2, \ldots, N$. In the similar way, average texture feature values of test image are computed. Then, the skin disease classification is done by computing Chi. square distance between train and test images.

4 Fuzzy Logic and GLCM-Based Skin Disease Classification

GLCM is implemented on all the training images, from which texture features of each and every training image are obtained. Then, the fuzzy logic (C-means clustering) is implemented on those values of the training image set. Fuzzy c-means (FCM) clustering is an iterative process. First, the initial fuzzy partition matrix is generated and the initial fuzzy cluster centers are calculated. In each step of the iteration, the cluster centers and the membership grade point are updated and the objective function is minimized to find the best location for the clusters. The process stops when the maximum number of iterations is reached [6]. In this way, cluster centers make a set of N values of each feature of one value which in turn reduces computational complexity drastically. Then, classification is done based on the similarity between cluster centers of training classes and the test image features.

5 Simulation Results

Images of skin diseases Angioedema, Actinic Keratosis and Eczema are collected from www.dermnet.com. They are cropped to 100 $\times$ 100 pixels size and converted to grayscale. Then, they are divided into training and test image sets.

5.1 *Implementation of Fuzzy Logic*

Firstly, Fuzzy logic is implemented on the class Angioedema, and the results are as shown in Fig. 3, where 1(bolded) represents the cluster center. From the Fig. 3, it can be observed that we get a single value for each feature. FCM clustering is applied on the other two classes Keratosis and Eczema and the results are given in Table 1.

5.2 *Skin Disease Classification*

Skin disease classification is done using Chi. square distance.

Chi Square distance, $d = \sum_{n=1}^{N} \frac{(S_n - M_m)^2}{S_n + M_m}; \; m = 1, \; 2, \; \ldots$

S_n, M_m, and N represent nth training image, mth test image and a number of training images, respectively. N is large for GLCM method where as N = 1 for the proposed method. The procedure is repeated for all the three classes. Then, the

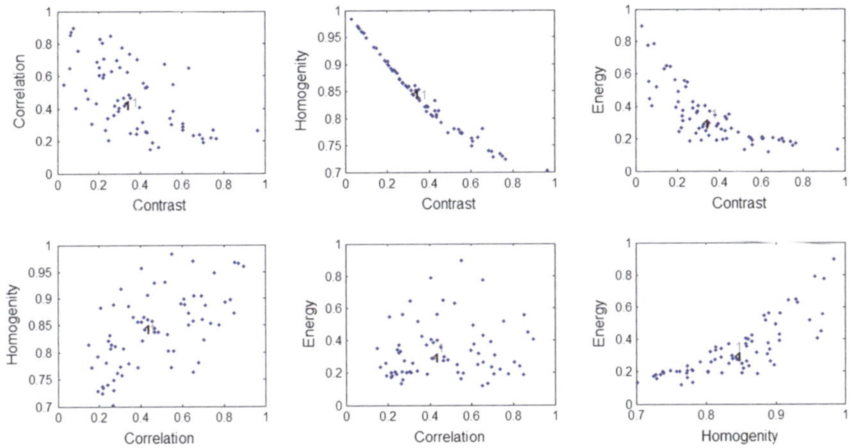

Fig. 3 Fuzzy C-means clustering on Angioedema

Table 1 Features of three classes of skin diseases after clustering

Skin disease class	Contrast	Correlation	Homogeneity	Energy
Angeoedema	0.35	0.46	0.84	0.30
Actinic Keratosis	0.40	0.58	0.85	0.34
Eczema	0.25	0.57	0.86	0.36

class, whose distance comes out to be minimum, is said to be the class the test image belongs to.

5.3 *Efficiency Calculation*

% Efficiency = 100−%Error where *%Error = No. of false detections/No. of test images.* In simulations, we obtain % Efficiency with GLCM method as 79% whereas % Efficiency with GLCM and Fuzzy c-means clustering method as 84%.

6 Conclusions

Three skin disease classes Angioedema, Actinic Keratosis, and Eczema are considered for skin disease classification. Texture features are extracted using normalized symmetrical GLCMs. Fuzzy c-means clustering is applied on the texture features of three classes of training data sets. Then, Chi. Square distance is used for image classification. Proposed GLCM and FCM method reduces the time consumed in skin disease classification and also, it exhibits high efficiency as compared to GLCM method.

References

1. Goyal, A.: Around 19 crore Indians likely to suffer from skin diseases by 2015—notes Frost & Sullivan, May 03 (2014)
2. Sparavigna, A., Marazzato, R.: An image processing analysis of skin textures. Skin Res. Technol. **16**(2), 161–167 (2010)
3. Al Abbadi, N.K., et al.: Psoriasis detection using skin color and texture features. J. Comput. Sci. **6**(6), 648–652 (2010)
4. Haralick, R.M.: Statistical and structural approaches to texture. Proc. IEEE **67**(5), 786–804 (1979)
5. Parekh, R.: Using texture analysis for medical diagnosis. Media med., IEEE Comput. Soc. **19**, 28–37 (2012)
6. Chattopadhyay, S.: A comparative study of fuzzy C-means algorithm and entropy-based fuzzy clustering algorithms. Comput. Inf. **30**, 701–720 (2011)

Melanoma Skin Cancer Detection Using Image Processing

Nishtha Garg, Vishakha Sharma and Prabhjot Kaur

Abstract Scientists have been trying to implement conventional ways across the world especially in developing and developed countries to cure the deadliest form of skin cancer in human which is known as Melanoma. But efforts are always blockaded by various challenges like high cost of sustaining traditional telemedicine and less availability of experts. There are broadly three types of skin cancer: basal cell cancer, squamous cell cancer, and melanoma. Greater than 90% of the cases are caused by exposure to ultraviolet radiation from the sun. It is important to detect cancer at the initial stage; only an expert dermatologist can classify which one is melanoma and which one is non-melanoma. A short time ago, there has been high implementation of techniques such as dermoscopy or epiluminescence light microscopy (ELM) in helping diagnosis. Using ELM is not affordable and objective, thus researchers motivated in automation diagnosis. This paper is intended to take a digital image, followed by preprocessing of the image to filter the extra noise present in the image. After this, skin lesion is subjected to segmentation and feature extraction with the implementation ABCD rule which will test the skin lesion on various parameters like asymmetry, border irregularity, color, and diameter of the lesion.

Keywords Melanoma · Segmentation · Feature extraction · ABCD rule

N. Garg (✉) · V. Sharma · P. Kaur
Maharaja Surajmal Institute of Technology, Janakapuri, Delhi, India
e-mail: nish.virgo103@gmail.com

V. Sharma
e-mail: sks.vishakha@gmail.com

P. Kaur
e-mail: prabhjot.kaur@msit.in

S. Urooj and J. Virmani (eds.), *Sensors and Image Processing*, Advances in Intelligent Systems and Computing 651, https://doi.org/10.1007/978-981-10-6614-6_12

1 Introduction

Melanoma is the most destructive type of skin cancer found in human body. It is formed from melanocytes cells. In women, most common site of melanoma is the legs and in men, it's the back. It is most common in the places which are highly exposed to sunlight like North America, Latin America, southern Africa. In this geographic pattern, the primary cause of melanoma is ultraviolet light exposure in association with the amount of skin pigmentation in the population. Melanin which is responsible for the color of the skin is produced by melanocytes which are prominently occured in skin but also found in bowel and eye. Melanoma can originate in any part of the body that contains Melanocytes [1].

If melanoma is detected at the initial stage, it can be treated with surgical eradication of the cancerous cells. The chances of melanoma occur again or spread depends on how deeply it has crux into the layers of the skin. In 2012, all over the world, 232,000 people were detected with melanoma, and 55,000 were resulted in death. Australia and New Zealand having the highest rate of melanoma detection found maximum death [1].

The majority of melanoma detection in earlier time is to analyze images using dermoscopy which is also called epiluminescence microscopy (ELM). In this technology, a dermatoscope is used to visually analyze the skin. A dermoscopic oil is applied on the cancerous skin to elucidate subsurfaces of the epidermis. This produces standardized lightning conditions and shows better view of the epidermis. One more conventional method of detecting melanoma is biopsy. It is very hurtful, less cost effective, and highly time-consuming technique [2].

So, to overcome all these challenges, this paper has incorporated digital image processing which makes it possible to detect cancer without touching the skin (Fig. 1).

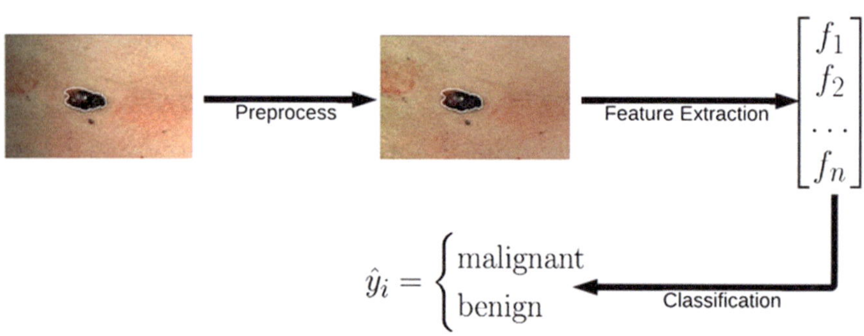

Fig. 1 Workflow of cancer detection support system [3]

2 Literature Review

Analysis of melanoma skin cancer was exercised from far back to ancient times. By observing the literature [2–8], it was found that many experiments are conducted on general skin cancer without any automatic approach. Researchers have proposed many techniques to overcome this effect by considering and working upon the images of skin cancer taken via various electronic devices. In the year 1999, Xu L. et al. have proposed a technique [2], wherein he has used various substages to improve the results. He firstly reduces a color image into an intensity image by intensity thresholding. Then, he used double thresholding to focus on an image area where a lesion boundary potentially exists. Further to detect the accurate boundary of a skin lesion, he used closed elastic curve technique. Robert Amelard et al. [3] have proposed a framework that performed illumination correction and feature extraction on skin images using high-level intuitive feature (HLIF). In the year 2013, N.S. Ramteke et al. have proposed a technique based upon ABCD rule to detect skin cancer. Within this technique, he used watershed segmentation technique for image segmentation, border detection, and decision as per the structural nature of the lesion. He further improved the image using wavelet transformation and implemented ABCD rule for cancer diagnosis. The limitation of this technique is that it did not implement 'E' factor of ABCD rule which can increase the accuracy of results [4]. In [5], the author has proposed a technique which detects malignant and benign melanoma skin cancer by extracting unique features through 2D wavelet transform. Then, the resultant image is given as input to artificial neural network classifier. But the limitation of the technique is it can detect results up to accuracy level of 84%.

3 ABCD Rule of Skin Cancer Detection

This section explains the ABCD rule of skin cancer detection [4].

3.1 Asymmetry (A)

Skin lesions are checked for symmetry and are divided into two parts. If both the halves are same, skin lesion is said to be non-cancerous; otherwise, it is said to be cancerous and possibly it can be the case of melanoma. Weighage for this factor is 1.3.

3.2 Border Irregularity (B)

B factor verifies the skin lesion edges. If the edges are jagged, serrated, or blurred, then skin lesion can be the case of melanoma. Value for border irregularity factor is based on 4 parameters, i.e., variance, parameter calculated from asymmetry, mean and variance of gradient of an image, and fractal of an image. Weighage for this factor is 0.1.

3.3 Color (C)

Pigmentation for skin lesion is not uniform in the case of cancerous lesions. To calculate the pigmentation non-uniformity, the presence of up to 6 colors must be detected—white, black, red, blue, dark brown, and light brown. The presence of each color adds 1 to the value of C factor. Weighage for this factor is 0.5.

3.4 Diameter (D)

Cancerous lesions are greater than 6 mm wide. Differential structures with at least five patterns are relevant for specific types of lesions. Any growth of a mole should be of concern. Diameter is calculated using the 'regionprops' function.

Total dermatoscopy score (TDS) index [4] is amalgamation of all the four factors describes above. All the factors are added with accordance to their weighage to find total score, which is used to distinguish the skin lesions to be cancerous (case of melanoma) or non-cancerous (case of non-melanoma) (Table. 1).

Formula for calculating TDS Index is as follows:

$$TDS = 1.3A + 0.1B + 0.5C + 0.5D \tag{1}$$

Table 1 Range of TDS

Range of TDS Index	Nature of skin cancer
TDS Index < 4.75	Non-cancerous skin lesion (Benign)
4.75 < TDS Index < 5.45	Suspicious area of skin lesion
TDS Index > 5.45	Cancerous skin lesion (Melanoma)

4 Proposed Approach

This paper has proposed a method which comprised of five steps. The steps are explained in the next sub-section. The paper has used MATLAB 7 [9] for implementation (Fig. 2).

4.1 Illumination Correction Preprocessing

The first step in preprocessing is illumination correction. The obtained image is in natural settings so illumination across the image is hidden by many factors like camera flash, office lightning, and natural lightning. It aims to adjust the original image pixel values to standardize lightening exposure across the complete image. This increases the decisiveness of the steps based on pixel values. For illumination correction, the proposed technique used three different algorithms sequentially (imfilter, imadjust, and median filter functions of MATLAB) (Fig. 3).

4.2 Skin Lesion Segmentation

The process of distinguishing image's pixel into semantic groups is called segmentation. Segmentation is aimed to get meaningful part from the image for

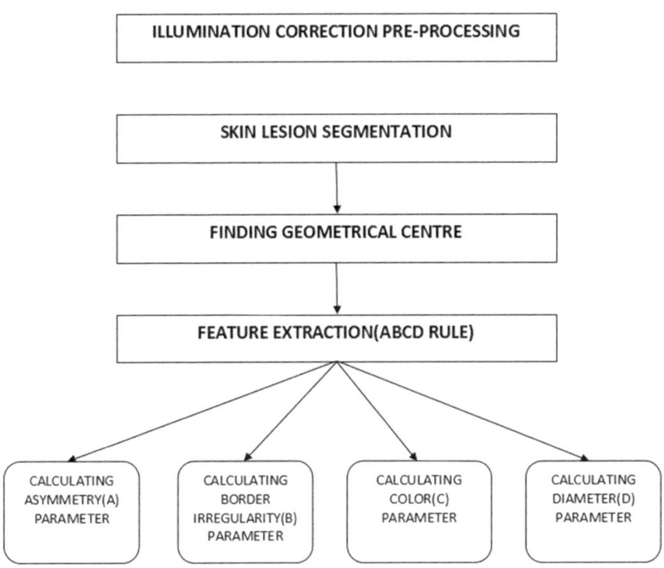

Fig. 2 Steps of proposed methodology

Fig. 3 Steps of illumination correction

Fig. 4 Skin lesion segmentation

simplification of image analysis. It determines the border that separates cancerous part from surrounding tissues. It results in binary mask outlining the skin lesion. The proposed technique has used thresholding approach to segment skin lesion by using Otsu method (Fig. 4).

4.3 Finding Geometrical Center

The purpose of this phase is to find the center of the segmented image. The purpose of the geometrical center of the segmented part is to verify the asymmetry property (A) of ABCD rule. It is also used to calculate diameter (D) of ABCD rule (Fig. 5).

Fig. 5 Finding geometrical center

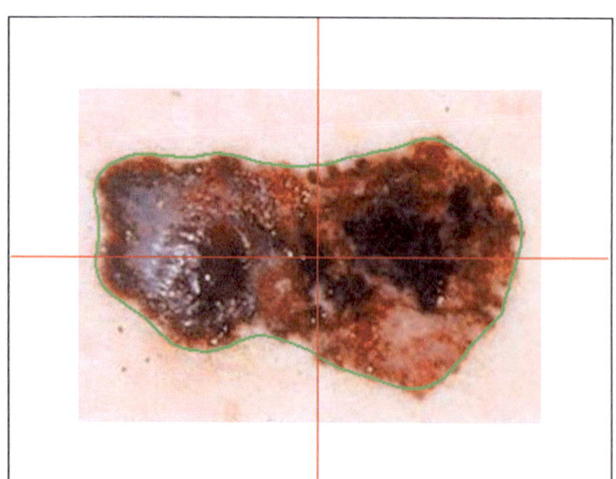

5 Result

The results have been evaluated on the data sets of skin images taken from dermis and medical image gallery [10]. Test images used in this paper are shown in Fig. 6. The results obtained by implementing proposed technique are recorded in Table 2, and the TDS calculations are listed in Table 3. Results are compared with the information mentioned on Mayo Clinic [11] and dermis.net [10]. Test images are shown in Fig. 6.

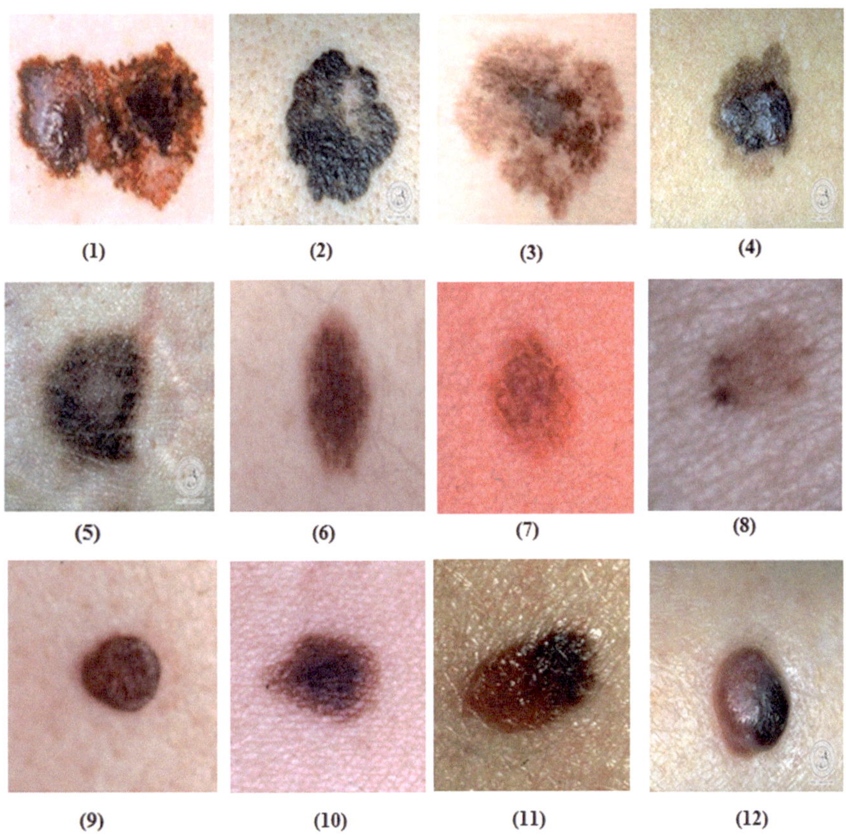

Fig. 6 Test images used in proposed technique

Table 2 Observed ABCD parameters of test images

Image no.	A factor	Weighted value	B factor	Weighted value	C factor	Weighted value	D factor	Weighted value
1.	0.476	0.62	3.6	0.36	5	2.5	4.18	2.09
2.	0.5	0.65	5.1	0.51	6	3	6.24	3.12
3.	0.48	0.63	5.8	0.58	5	2.5	3.98	1.99
4.	0.48	0.63	7.3	0.73	6	3	5.4	2.7
5.	0.5	0.65	7.9	0.79	6	3	5.44	2.72
6.	0.5	0.66	2.8	0.28	3	1.5	1.8	0.90
7.	0.49	0.64	3.5	0.35	3	1.5	1.86	0.93
8.	0.49	0.64	4.9	0.49	4	2	1.84	0.92
9.	0.5	0.65	2.4	0.24	3	1.5	1.82	0.91
10.	0.49	0.64	3.9	0.39	5	2.5	1.84	0.92
11.	0.37	0.49	6.1	0.61	4	2	4.6	2.3
12.	0.5	0.65	3.8	0.38	6	3	5.1	2.55

Table 3 Comparison of proposed algorithm result with actual results

Image no.	**TDS**	Proposed algorithm result	Actual results	Correctness
1.	5.58	Melanoma	Melanoma	Correct
2.	7.30	Melanoma	Melanoma	Correct
3.	5.71	Melanoma	Melanoma	Correct
4.	7.35	Melanoma	Melanoma	Correct
5.	7.37	Melanoma	Melanoma	Correct
6.	3.38	Benign	Benign	Correct
7.	3.44	Benign	Benign	Correct
8.	4.06	Benign	Benign	Correct
9.	3.31	Benign	Benign	Correct
10.	4.4	Benign	Benign	Correct
11.	5.4	Suspicious	Benign	Incorrect
12.	6.58	Benign	Benign	Correct

6 Conclusion and Future Scope

The diagnosis of skin lesion has been implemented on complete data set of images. It is found to be working excellently with the accuracy up to 91.6%. Automatic diagnosis is very simple, feasible, and achievable with the well-defined implementation of segmentation, feature extraction, and classification. Further, the accuracy could be improved with the implementation of E factor with ABCD rule. Also, other parameters of dependency can also identified. Illumination correction of the image can be improved for more noise removal.

References

1. http://en.wikipedia.org/wiki/Melanoma
2. Xu, L., et al.: Segmentation of Skin cancer images. Elsevier Image Vis. Comput. **17**, 65–74 (1999)
3. Amelard, R., Glaister, J., Wongand, A., Clausi, D.A.: Melanoma decision support using lightening—corrected intuitive feature model. Computer vision techniques for the diagnosis of skin cancer, pp. 193–219. Springer, Berlin (2014)
4. Ramteke, N., Ramteke, N.S., Jainz, S.V.: ABCD rule based automatic computer-aided skin cancer detection using MATLAB. Int. J. Comput. Technol. Appl. **4**(4), 691–697 (2013)
5. Asvin, R., Jaleel, J.A., Salims, S.: Implementation of ANN classifier using MATLAB for skin cancer detection. Int. J. Comput. Sci. Mob. Comput. **3**(5), 87–94 (2013)
6. Ganster, H., et al.: Automated Melanoma recognition. IEEE Trans. Med. Imaging **20**(3), 233–239 (2001)
7. Sigurdsson, S.: Detection of Skin cancer by classification of Raman Spectra. IEEE Trans. Biomed. Eng. **51**(10), 1784–1793 (2004)
8. Okuboyejo, D.A.: Automating skin disease diagnosis using image classification. In Proceedings of the World Congress on Engineering and Computer Science. II, WCESC 2013, 23–25 (2013)
9. MATLAB version 7.10.0. Natick, Massachusetts: The MathWorks Inc. (2010)
10. http://www.dermis.net and http://www.dermquest.com
11. www.mayoclinic.org

Performance Assessment for Clustering Techniques for Image Segmentation

Jyoti Arora, Kiran Khatter and Meena Tushir

Abstract The analysis and processing of large datasets is a challenge for researchers. Several approaches have been developed to model these complex data, based on some mathematical theories. In this paper, we are comparing image segmentation techniques based on unsupervised clustering approaches such as probability approach, possibilistic approach, credibilistic, and intuitionistic approach. This paper presents comparison of these approaches. The four approaches are studied and analyzed both quantitatively and qualitatively. These approaches are implemented on synthetic images, and performance is analyzed.

Keywords FCM · Clustering · Possibilistic c-means · Intuitionistic c-means · Credibilistic c-means

1 Introduction

Image segmentation plays an important part in image analysis and computer vision. It is used in different application areas including object detection, medical imaging, biometrics, and pattern recognition. Segmentation process divides the image into desired parts separating the regions from image on the basis of intensity, color, texture, or other features of each pixel. Meaningful segmentation is the first step

J. Arora (✉)
Department of Information Technology, Maharaja Surajmal Institute of Technology, Delhi, India
e-mail: joy.arora@gmail.com

K. Khatter
Department of Computer Science, Ansal University, Gurgaon, India
e-mail: kirankhatter@ansaluniversity.edu.in

M. Tushir
Department of Electrical & Electronics Engineering, Maharaja Surajmal Institute of Technology, Delhi, India
e-mail: meenatushir@yahoo.com

S. Urooj and J. Virmani (eds.), *Sensors and Image Processing*, Advances in Intelligent Systems and Computing 651, https://doi.org/10.1007/978-981-10-6614-6_13

from low-level image processing, transforming a gray scale or color image into one or more other images to high-level image description in terms of objects, features, and scenes. These images are very complex data; an accurate partitioning of an image is very challenging problem. Till now various techniques have been used for segmentation of image. According to the literature [1–3], image segmentation techniques can be broadly divided into four categories: thresholding, region extraction, edge detection, and clustering. In the processing of an image, clustering and segmentation are two terms considered close to each other. Clustering refers to analyze the clusters in an image on the basis of color information which can be visualized in an image, and mapping the clusters onto spatial domain by physically separating the regions on the basis of color information is called segmentation. In this paper, fuzzy clustering-based image segmentation techniques are considered.

The fuzzy set theory was given by Zadeh [4] in 1865 with a view to reconcile mathematical relation with human knowledge and engineering sciences. The most widely used fuzzy algorithm is fuzzy c-means (FCM) given by Bezdek [5, 6] where every data point is defined by membership grade between 0 and 1.

In this paper, we are comparing four fuzzy-based image segmentation techniques namely FCM Clustering, Possibilistic c-means, Intuitionistic Fuzzy c-means, Credibilistic C-Means for digital image with the effect of noise. The remainder of paper is sectioned as follows. Section 2 provides related work which includes detail of all the four techniques, Sect. 3 compares these four techniques by inducing different level of noise with these images and compares the result quantitatively and qualitatively. Section 4 concludes the results.

2 Related Work

2.1 *Fuzzy C-Means Algorithm*

Fuzzy c-means algorithm (FCM) is the most widely used algorithm. This approach partitions a set of data $\{x1,\ldots\ldots\ldots\ldots\ldots, xn\} \subset R^s$ into C clusters based on a similarity computed by the least square function of Euclidean distance metrics. Let X denotes the image data which is divided into C clusters depending on the intensity information of image. The objective function of FCM is defined by Bezdek [6] as

$$J_{FCM}(X:U,\ V) = \sum_{j=1}^{c}\sum_{i=1}^{N}(u_{ij})^m\left\|x_i - v_j\right\|^2,\ 1<m<\infty \tag{1}$$

where m is always greater than 1 and is a fuzzification parameter which determines the rate of decay of membership value, $v_j \in R^S$ is a cluster center, $u_{ij} \in [0,\ 1]$ is a degree to which data x belongs to cluster, defines partition matrix. $\left\|x_i - v_j\right\|^2$ is a Euclidean distance metrics. The partition matrix U is a convenient tool for representing cluster structure in the data, where the fuzzy partition has constraint $\sum_{i=1}^{c} u_{ij} = 1$ which states that the membership degree of a data point across all cluster sums to 1.

2.2 Possibilistic C-Means Algorithm

In 1996, Krishnapuram and Keller [7] given the possibilistic approach of fuzzy clustering in order to overcome the drawbacks of FCM. PCM is a mode seeking algorithm. In the PCM, clusters formed are more dense. The objective function of PCM includes the typicality which defines the "zone of influence" of the point. The objective function of PCM is defined as

$$J_{PCM}(X : U,\ V) = \sum_{j=1}^{c} \sum_{i=1}^{N} (u_{ij})^m \|x_i - v_j\|^2 + \eta_i \sum_{j=1}^{N} (1 - u_{ij})^m \quad (2)$$

The membership matrix U generated by the PCM is not a partition matrix since it does not satisfy the membership constraint of FCM. The parameter η_i is a typicality, u_{ij} is a membership, and $v_j$is a center referred from [7, 8].

Here, minimization of the objective function is done by alternating optimization. The distinguishing feature of PCM is that membership u_{ij} of any data point is not relative and depends on the distance between data point to the center, thus the centers do not affect each other during optimization of objective function of PCM. This often leads to the problem of overlapping clusters. Also, a PCM result relies on the initialization of the algorithm with initial center and membership matrix. Generally, FCM is used to initialize PCM. However, if the outlier is distant, PCM will not be able to detect due to poor initialization from FCM.

2.3 Credibilistic Fuzzy C-Means

To minimize the effect of outliers, Krishna K. Chintalapudi [9] proposed credibilistic fuzzy c-means (CFCM) and introduced a new variable i.e., credibility. CFCM defines credibility as:

$$\psi_k = 1 - \frac{(1-\theta)\alpha_k}{\max_{j=1..n}(\alpha_j)},\ 0 \le \beta \le 1 \ \text{ where } \ \alpha_k = i = 1.....c^{\min}(d_{ik}) \quad (3)$$

setting, the value of $\theta = 1$ reduces the scheme to FCM while $\theta = 0$ assigns zero membership to the most noisy vector. If $\theta = 1$, then there is no noisy vector present in the dataset. Thus, we choose $\theta = 0$ in all our implementation so as to separate noise from the dataset.

CFCM partitions X image by minimizing the objective function of FCM as represented in Eq. (1) with the following constraints applied

$$\sum_{i=1}^{i=c} u_{ik} = \psi_k;\ k = 1\ldots\ldots\ldots n. \tag{4}$$

The condition for membership is given by Eq. (5)

$$u_{ik} = \frac{\psi_k}{\sum_{j=1}^{j=c} \frac{d_{ik}}{d_{jk}}^{\frac{2}{m-1}}} \forall i,\ k \tag{5}$$

Memberships generated by CFCM for outliers are lower than those generated by FCM because for outliers, credibility is very small. The main advantage of CFCM is reducing the effect of outliers on regular clusters.

2.4 *Intuitionistic Fuzzy C-Means*

Intuitionistic fuzzy c-means [10] works on generalized fuzzy sets in which elements are characterized by both characteristics of membership and non-membership values. Atanassov introduced a parameter called hesitation degree $\pi_A(x)$ which explains lack of knowledge in defining the membership degree of all elements x in the set A. It is calculated as follows:

$$\pi_A(x) = 1 - \mu_A(x) - v_A(x) \tag{6}$$

where $\mu_A(x)$ is the membership degree, $v_A(x)$ is the non-membership degree, and $\pi_A(x)$ is the hesitation degree. The objective function for intuitionistic fuzzy c-means (7) includes modified objective function of FCM and intuitionistic fuzzy entropy.

$$J_{IFCM} = \sum_{i=1}^{n} \sum_{k-1}^{c} u_{ik}^{*m} d_{ik}^2 + \sum_{i=1}^{n} \pi_i^* e^{1-\Pi_i^*} \tag{7}$$

$u_{ik}^* = u_{ik} + \pi_{ik}$, in which u_{ik}^* denotes the intuitionistic fuzzy membership, u_{ik} gives the normal fuzzy membership of the kth data in i^{th} class. From the above equation, π_{ik} denotes the hesitation degree, which is

$$\pi_{ik} = 1 - u_{ik} - \left(1 - u_{ik}^{\alpha}\right)^{1/\alpha},\ \alpha > 0 \tag{8}$$

Yager gave an intuitionistic fuzzy complement which is used to calculate above defined constant.

Intuitionistic fuzzy entropy (IFE) is the second term of objective function for IFCM given by Eq. (9).

$$IFE(A) = \sum_{i=1}^{n} \pi_A(x_i) e^{[1 - \pi_A(x_i)]} \tag{9}$$

As Euclidian distance measure is used in IFCM, hence only hyper-spherical clusters can be detected in the data [11]. Nonlinearly separable data can't be worked upon by IFCM.

3 Comparison and Experimental Results

The purpose of the experimental part was to segment the synthetic image of coins into different clusters based on the intensity value of the image using four clustering techniques i.e., FCM, PCM, CFCM, and IFCM. We will check the result of segmentation with number of clusters noc = 2, m = 2, $\varepsilon = 0.0002$, and maximum number of iterations defined are k = 100. We test the performance of different algorithms with the presence of noise. We have added salt and pepper noise with different levels of 0.02 and 0.04 to analyze the robustness of different techniques with noise on image dataset as shown in Fig. 1.

3.1 Qualitative Analysis

A synthetic image of coins (246 X 300) consisting of different coins in a gray scale is used as shown in Fig. 1. In the second row of Fig. 1d, e, f shows the result of FCM algorithm, we observe the segmentation of image into two clusters are well-defined with every coin is well separated in all the three cases but FCM is not able to remove noise effect which was added by salt and pepper noise. Third row of the Fig. 1g, h, i shows the result of PCM algorithm, we observe that good result of segmentation is shown by PCM, with the one coin in the center showing different display as compared to FCM. Noise effects are not completely separated out in both the cases. Fourth row of Fig. 1j, k, l shows the result of CFCM. It shows better result in the case of noisy images. It removes the noise to some extent as compared to other techniques. Figure 1m, n, o shows the result of IFCM. In this case, it also shows good results in case segmentation, and effect of noise is also reduced, while observing inside the coin area which is showing more clarity as compared to other

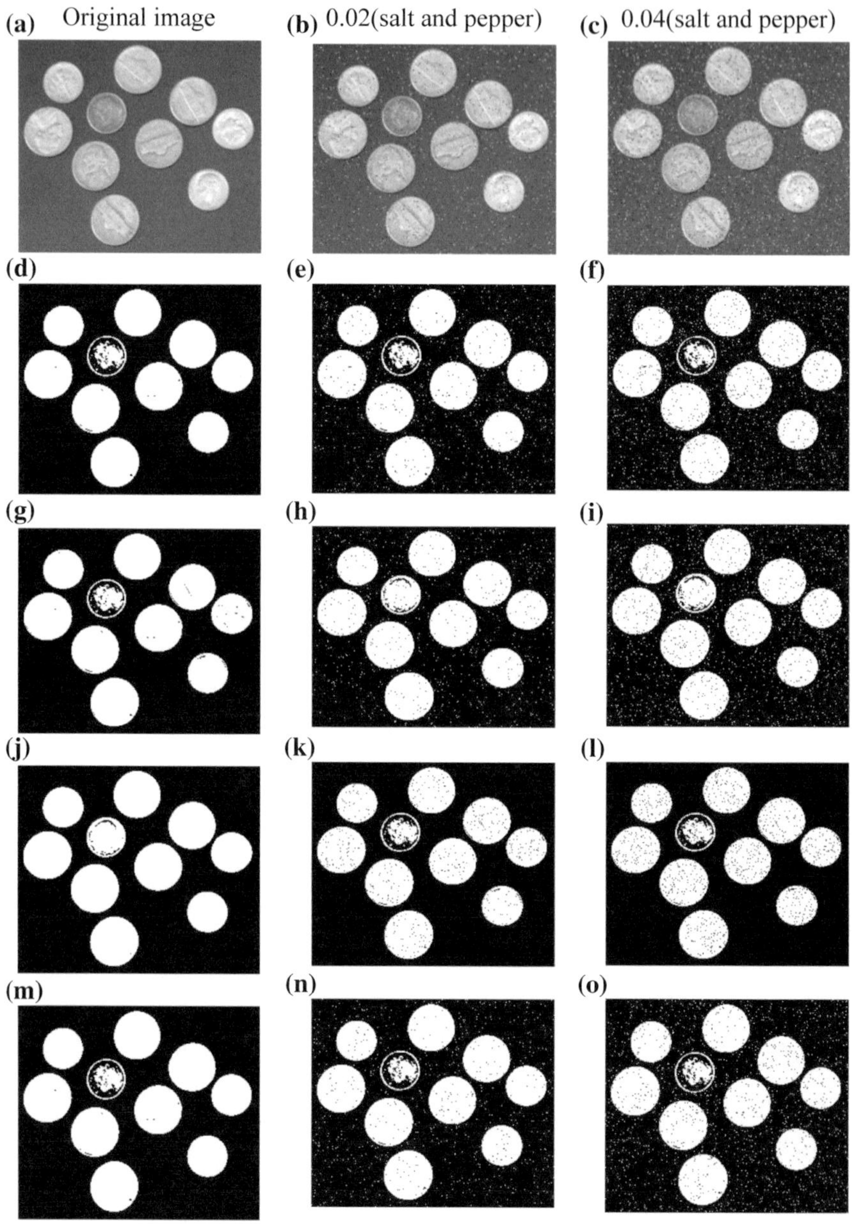

Fig. 1 **a** represents the original image; **b** image distorted with 0.02 salt and pepper noise; **c** image distorted with 0.04 salt and pepper noise; **d**, **e,** and **f** are output of FCM; **g**, **h,** and **i** are output of PCM, **j**, **k,** and **l** are output of CFCM; **m**, **n,** and **o** are output of IFCM

techniques. So from these results, we can analyze that both FCM and PCM show equivalent results if our image is noisy. CFCM and IFCM are showing better results as compared to other two techniques even in the presence of noise.

3.2 *Quantitative Analysis*

Quantitatively, we have compared these algorithms by analyzing on the basis of time complexity of each algorithm i.e., execution time of each technique to segment the image data in the absence and presence of noise. We also calculated that a number of iterations were used to segment an image. Table 1 shows the execution time taken by each algorithm to achieve its local minima which are in seconds and the value of k (number of iterations) for which this value of minima is attained. From Table 1, we can analyze that time complexity of IFCM is more as compare to other techniques. Number of iterations used by IFCM is more as compared to other algorithms. CFCM with good initialization of FCM also takes optimal number of iterations to compute the result. PCM takes least number of iterations although it was initialized by FCM center and membership matrix. FCM takes optimal number of iterations and execution time to calculate the results. So, from these results, it can be analyzed that IFCM and CFCM with the introduction of new parameters are more complex as compared to FCM and PCM but show better results in the presence of noise. Comparatively, the robustness of algorithm increases the time complexity of that algorithm.

Table 1 Performance comparison of FCM, PCM, CFCM, and IFCM in terms of execution time and number of iterations

Technique	Image	No. of iterations	Execution time (s)
FCM	Noiseless	11	0.225
	0.02	12	0.233
	0.04	12	0.363
PCM	Noiseless	3	0.339
	0.02	3	0.299
	0.04	4	0.311
CFCM	Noiseless	9	0.437
	0.02	9	0.417
	0.04	10	0.452
IFCM	Noiseless	23	0.541
	0.02	21	0.498
	0.04	20	0.445

3.3 Validity Index

Validity Index is used to evaluate the number of segments into which an image is to be segmented. They are used to measure the segment accuracy. Here, we have used Validity Partition Coefficient [12] to evaluate the number of segments. Best partition results are achieved when the Vpc is maximum. It has been observed that when n = 2, V_{pc} is maximum, and it decreases as n increases. The square symbol represents the value of Vpc at particular value of n. Hence, in each clustering techniques, the number of clusters we have given as input for segmentation is 2. The function of Vpc is given by Eq. (10).

$$V_{pc} = \sum_{i=1}^{n} \sum_{k=1}^{c} u_{ik}^{2} \tag{10}$$

4 Conclusion

Digital image considerable contains unknown noise and uncertainity in data. FCM, which is widely used as fuzzy clustering algorithm is intensity-based algorithm which is not robust against noisy images. Upon carefully examining the results, we observe that PCM segments the image and generates good result in the absence of noise, but it fails to show the good results in the presence of noise. In case of presence of noise, credibilistic fuzzy c-means shows better result as compared to FCM and PCM with the increase in computational time and number of iteration used to converge to minimum. Intuitionistic fuzzy c-means also gives good results as compared to FCM and PCM but the computational time taken by IFCM is maximum among all four algorithms. Thus adding some penalty term to FCM which can separate segmentation techniques based on some different features of image as entropy, contour and thresholding etc. which can preprocess image, so that segmentation of preprocess data can improve the quality of segmentation. In the future, we would like to work on some combination of these techniques.

References

1. Chandrashekhar, A.M., Raghuveer, K.: Performance evaluation of data clustering technique using KDD Cup-99 intrusion detection data set. Int. J. Inf. Netw. Secur. **1**(4), 294–304 (2012)
2. Shih, C.K., Hong-Long, T., Sharon, C.: Fuzzy C-means clustering with spatial information for image segmentation. Comput. Med. Imaging Graph. **30**, 2–15 (2006)
3. MymoonZuviria, N., Deepa, M.: A robust fuzzy neighborhood based C means algorithm for image clustering. Int. J. Adv. Res. Comput. Sci. Softw. Eng. **3**(3), 87–94 (2013)
4. Zadeh, L.A., Deepa, M.: Fuzzy sets. Inf. Control **8**, 338–353 (1965)

5. Bezdek, J.C.: Pattern recognition with fuzzy objective function algorithms. Plenum, New York (1981)
6. Bezdek, J.C., Pal, S.K.: Fuzzy models for pattern recognition. IEEE, New York (1981)
7. Krishnapuram, R., Keller, J.M.: A possibilistic approach to clustering. IEEE Trans. Fuzzy Syst. **1**(2), 98–110 (1993)
8. Dave, R.N.: Characterization and detection of noise in clustering. Pattern Recognit. Lett. **12**, 657–664 (1991)
9. Chintalapudi, K.K., Kam, M.: A noise resistant fuzzy c-means algorithm for clustering. IEEE Conf. Fuzzy Sys. Proc. **2**, 1458–1463 (1998)
10. Chaira, T.: A novel intuitionistic fuzzy c means clustering algorithm and its application to medical images. Appl. Soft Comput. **11**, 1711–1717 (2011)
11. Kaur, P., Soni, A.K., Gosain, A.: Novel intuitionistic fuzzy c-means clustering for linearly and nonlinearly separable data. WSEAS Trans. Comput. **11**(3), 2224–2872 (2012)
12. Xie, X.L., Beni, G.A.: Validity measure for fuzzy clustering. IEEE Trans. Pattern Anal. Mach. Intell. **3**, 841–846 (1991)

Automatic Seed Point Selection in B-Mode Breast Ultrasound Images

Madan Lal and Lakhwinder Kaur

Abstract In order to reduce the processing time and complexity to detect the boundary of lesions in breast ultrasound (BUS) images, first step is selection of region of interest (ROI), which subsequently needs selection of seed point. Seed point is starting point that lies inside the lesion region. After selection of seed point, region growing techniques are used for segmentation of lesions or for selection of region of interest. Seed point can be selected manually, but it needs human interaction. To design a fully automatic breast ultrasound computer-aided diagnosis (CAD) system, an automatic seed point selection technique is required. In this paper, an automatic seed point detection technique is proposed. This technique is applied on 108 BUS images (57 benign and 51 malignant). Results are compared with other available methods. Quantitative experiment results show that this method could find the proper seed point for 95.3% BUS Images.

Keywords Seed point · Region of interest (ROI) · Region growing · Breast ultrasound (BUS) images

1 Introduction

Sonography is a most commonly used technique for imaging abnormalities in human body [1–3]. Due to its noninvasive nature, nowadays it is widely used for breast cancer detection [4]. However, due to its nature BUS (breast ultrasound)

50th Golden Jubilee Annual Convention,
CSI-02nd–05th December 2015, India International Centre, New Delhi

Madan Lal (✉) · Lakhwinder Kaur
Department of Computer Engineering, Punjabi University, Patiala, India
e-mail: madanlal@pbi.ac.in

Lakhwinder Kaur
e-mail: mahal2k8@yahoo.com

S. Urooj and J. Virmani (eds.), *Sensors and Image Processing*, Advances in Intelligent Systems and Computing 651, https://doi.org/10.1007/978-981-10-6614-6_14

images suffer from low contrast, shadow effect, and speckle noise [5]. Such parameters make it difficult to directly identify the abnormalities in BUS images. In recent years, some methods have been suggested to detect the exact shapes of abnormalities in BUS images [6–8]. In most of these techniques, lesions are segmented using region growing methods whose basic requirement is the selection of initial seed. Seed point can be selected manually, but it needs human interaction which is time-consuming. To make the system independent, some automatic seed point selection techniques are proposed.

2 Related Work

As the performance of region growing segmentation methods heavily depends on the selection of initial accurate seed selection, significant work has been done in this area [6–9]. In [6], image is first enhanced by using some preprocessing techniques, and then, a quantitative method is applied to calculate a set of some points. Highest scoring point in the set is selected as the seed point. In this method, it is assumed that the lesions usually appear in the middle of the image, which is not always true. In [7], input image is preprocessed and morphological operations are performed; then, a binary image is obtained. In binary image, the sum of the pixels in each column and row are calculated. The row and column numbers with the max sums, respectively, are used as indexes of seed point. In [8], seed point and threshold (which control the region growing process) are automatically calculated using the texture features of the image. Energy and entropy parameters of co-occurrence matrix are used to find the co-occurrence features which are further used to distinguish homogeneous and nonhomogeneous regions. Then, the run length features are calculated for selected points. A point is considered as seed point if the run length features of selected point and its neighboring points are equal. In [9], after preprocessing image is converted into binary image by using iterative threshold selection method, i.e., after binarization, all boundary-connected regions which are not connected with a half-size co-centered window of the original image are discarded. All the remaining regions are ranked using a mathematical formula, and the center of region having highest rank is considered as a seed point.

In the existing methods, mostly consider only the texture features of abnormal regions and only a few consider the spatial features. In this paper, a new technique is suggested, which uses both texture features and spatial features of an input B-Mode BUS to find the automatic seed point.

3 Proposed Method

The following steps are used to implement the proposed technique.

3.1 Image Enhancement

BUS has inherent artifacts like speckle and low contrast. For accurate seed point selection, it must be preprocessed. In this work, speckle reducing anisotropic diffusion (SRAD) [10] technique is used. It is a de-speckle method used to reduce speckle noise and preserve the edges. SRAD uses adaptive weighted filters and processes the noisy image iteratively. The diffusion coefficient is determined by Eq. (1).

$$C(q) = \frac{1}{\left[1 + q^2(x,\ y;\ t) - q_0^2(t)\right] / \left[q_0^2(t)\left(1 + q_0^2(t)\right)\right]}. \tag{1}$$

And the instantaneous coefficient is given by Eq. (2)

$$q(x,\ y;\ t) = \sqrt{\frac{\left(\frac{1}{2}\right)\left(\frac{|\nabla I|}{I}\right)^2 - \left(\frac{1}{4^2}\right)\left(\frac{\nabla^2 I}{I}\right)^2}{1 + \left(\frac{1}{4}\right)\left(\frac{\nabla^2 I}{I}\right)^2}}. \tag{2}$$

The initial $q_0(t)$ called speckle-scaled function, which effectively controls the amount of smoothing applied to the image by SRAD, is given by Eq. (3).

$$q_0(t) = \frac{\sqrt{\mathrm{var}[Z(t)]}}{\overline{z(t)}}, \tag{3}$$

where $\sqrt{\mathrm{var}[z(t)]}$ and $\overline{z(t)}$ are the intensity variance and mean over a homogeneous area at t, respectively. Here, $q(x,\ y;\ t)$ serves as the edge detector in the speckled imaginary. The function exhibits high values at edges and provides low values in homogeneous regions where $z(t)$ is the most homogeneous area at t. In this experiment, iteration time used is 20.

3.2 Image Binarization

Enhanced B-Mode grayscale image is converted into black-and-white image by using average gray-level value as a threshold. After binarization, morphological operations are applied to clear the small white spots in black-and-white image.

Now for all the rows and columns, those have intensity values zero for all pixels in binary image are set to one in enhanced image. This step discards the darker regions that are near the boundary of an image. After setting the row and column pixel values to one, only the part of enhanced image which lies inswhich has minimum average intensity value amoide these white rows and columns is used for further processing.

3.3 *Image Subdivision*

In this step, image (output of step 3.2) is divided into different blocks of same size independent of the size of input image. Further processing is performed on the individual blocks. If there is an image of size $M \times N$ and we want to divide it into blocks of size $m \times n$, then there will be total $(m/m \times N/n)$ blocks where $M > m$ and $N > n$. In this work, image is divided into 100 blocks.

3.4 *Texture Feature Extraction*

Texture features of all the sub-blocks of segmented image are extracted by using gray-level co-occurrence matrix (GLCM) proposed by Haralick et al. [11]. GLCM computes the probability density function of image $f(x, y)$ for all pair of pixels (i) and (j) with distance (d) and with angular displacement $(\theta) = 0$, 45, 90, and 135 degree. This calculation computes the frequency of gray tone occurrence for angular adjacent pixels [12]. It reflects some quantitative features of the segment. In this work, values of these features are computed at directional parameter $\theta = 0$ degree and at distance parameter $d = 1$ (Fig. 1).

GLCM is able to find various features like homogeneity, energy, entropy, contrast, variance and correlation [12] of different segments of image, but in this work, only homogeneity parameter is used because homogeneity is the property of lesion region in BUS image [8, 13]. The segment for which value of homogeneity is comparatively high is considered as target segment of the image, provided it does not lie on the boundary of image.

Fig. 1 (1) A 4 × 4 image with four gray levels and corresponding co-occurrence matrix [8]

Image

0	0	1	1
0	0	1	1
0	2	2	2
2	2	3	3

Corresponding GLCM

$$\begin{bmatrix} 4 & 2 & 1 & 0 \\ 2 & 4 & 0 & 0 \\ 1 & 0 & 6 & 1 \\ 0 & 0 & 1 & 2 \end{bmatrix}$$

3.5 Target Segment Detection

One segment is selected that has minimum average intensity value among all the segments of image, provided it does not lie on the boundary of image. Another segment is selected that has second minimum average intensity value among all the segments with condition that it also should not lay on the boundary of image. Now homogeneity (calculated using gray-level co-occurrence matrix) of these segments is compared. The segment that is more homogeneous is selected. This segment of the image is considered as a target segment which contains the seed point.

3.6 Seed Point Selection

Based on the texture feature calculation, once the target block is selected its center is considered as an initial seed point. Now, intensity value of initially selected seed is compared with the intensity values of all the pixels in the target segment. The pixel that has minimum intensity value among all the pixels in this selected segment is considered as final seed point. If more than one pixel have same minimum intensity value, then the pixel that is near to center of target segment is considered as final seed point.

4 Proposed Algorithm

The following steps are used to find automatic seed point.

- Input B-Mode BUS image.
- De-speckle the input image using SRAD filter.
- Binarize the image using average intensity value of input image.
- Apply morphological operations to remove the small spots.
- Set all rows and columns whose all pixel values are zero in binarized image to one in enhanced image.
- Select the part of image which lies within white rows and columns. (Let its size be $M \times N$.)
- Divide the image into segments of size $m \times n$ where $m < M$ and $n < N$.
- Find the homogeneity of all the segments using GLCM.
- Select the segment which has minimum average intensity value among all the segments, provided it does not lie on the boundary of image. (Let it be segment A.)
- Select the next segment which has second minimum average intensity value with the condition that it also should not lie on the boundary of image. (Let it be segment B.)
- Select the segment as a target segment which is more homogeneous between A and B.

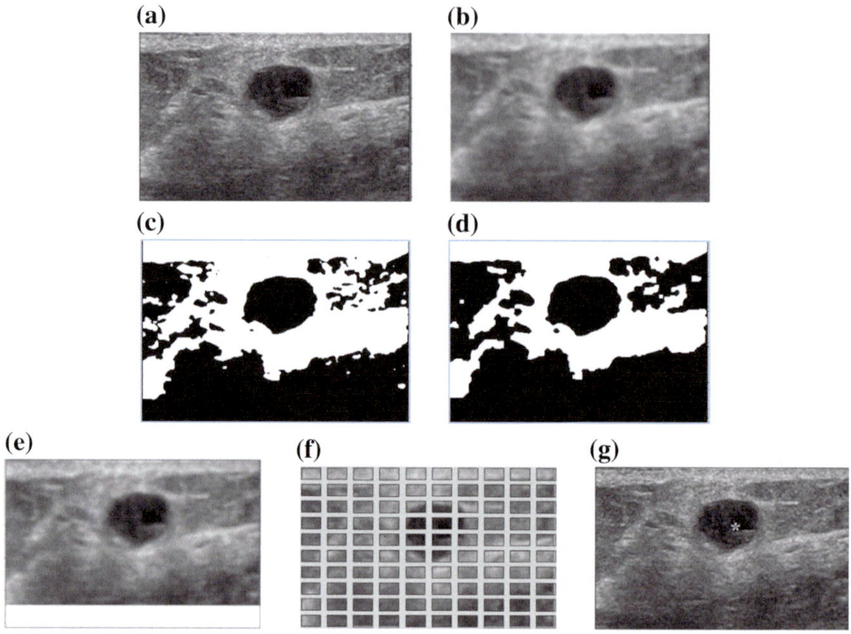

Fig. 2 **a** Original image. **b** Preprocessed image. **c** Binarized image. **d** Image after morphological operations. **e** Image after discarding rows and columns whose all pixel values are more than average intensity value of input image. **f** Image after segmentation. **g** Image showing final seed point

- If there is only pixel that has minimum intensity value among all the pixels in the target segment, then select it as final seed point; otherwise, consider the pixel that has minimum intensity value and also that is near to the center of target segment as a final seed point.

Figure 2 shows the output of the proposed technique at different stages. The image used in this experiment is taken from [14].

5 Experiment, Results, and Analysis

The proposed method is implemented by using MATLAB (R2012a). To verify the performance of this method, it is experimented with 108 B-Mode BUS images dataset (57 benign and 51 malignant). The proposed algorithm gives correct seed points (which lies inside the tumor region) for 103 images. This technique is also applied on images used in [9] to compare this method with Madabhushi's method. The results are displayed in Fig. 3 which shows that this method gives correct seed points for two different images whereas Madabhushi's method gives inaccurate

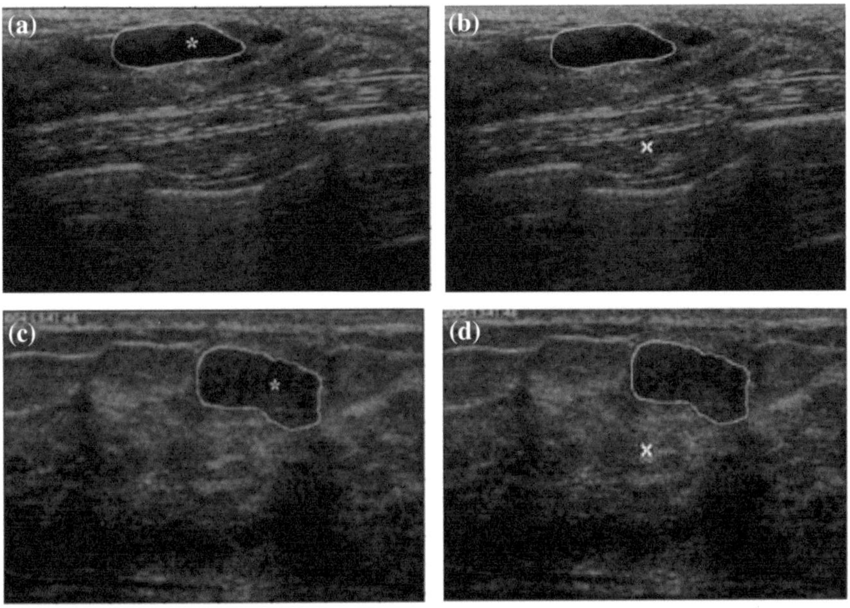

Fig. 3 **a** and **c** Results of the proposed method **b** and **d** the results of Madabhushi's method [9]

Table 1 Performance of the proposed method

Total input images	# Image for which seed point is detected correctly	# Image for which inaccurate seed point is detected
108	103, (95.37%)	5, (4.63%)

results. For image dataset, 68 images are collected from Department of Radiology, Government Rajindra Hospital, Patiala, Punjab, and 40 images are collected from different Web sites [14, 15] (Table 1).

6 Conclusion

In this paper, a completely automatic seed point selection method for B-Mode breast ultrasound images is presented. This technique uses both spatial and texture information of input image for calculation of initial point. The proposed algorithm is fast and its accuracy rate is 95.37%. It is also able to find accurate seed point for images which have shadow effect. For some malignant cases, it gives inaccurate seed point, so future work will be focused on to improve the same.

References

1. Virmani, J., Kumar, V., Kalra, N., Khandelwal, N.: SVM-based characterization of liver ultrasound images using wavelet packet texture descriptors. J. Digital Imaging **26**(3), 530–543 (2013)
2. Virmani, J., Kumar, V., Kalra, N., Khandelwal, N.: Characterization of primary and secondary malignant liver lesions from B-mode ultrasound. J. Digital Imaging **26**(6), 1058–1070 (2013)
3. Virmani, J., Kumar, V., Kalra, N., Khandelwal, N.: A comparative study of computer-aided classification systems for focal hepatic lesions from B-mode ultrasound. J. Med. Eng. Technol. **37**(4), 292–306 (2013)
4. Costantini, M., Belli, P., Lombardi, R., Franceschini, G., Mule, A., Bonomo, L.: Use of the sonographic breast imaging reporting and data system lexicon. Am. Inst. Ultrasound Med. **25**, 649–659 (2006)
5. Rolf, A., Leanne, B.: Seeded region growing. IEEE Trans. Image process. **16**(June), 641–647 (1994)
6. Madabhushi, A., Metaxas, D.N.: Combining low high-level and empirical domain knowledge for automated segmentation of ultrasonic breast lesions. IEEE Trans. Med. Imaging **22**(2), 155–169 (2003)
7. Jung, I.S., Thapa, D., Wang, G.N.: Automatic segmentation and diagnosis of breast lesions using morphology method based on ultrasound. Fuzzy Syst. Knowl. Discov. **3614**, 1079–1088 (2005)
8. Poonguzhali, S., Ravindran, G.: A complete automatic region growing method for segmentation of masses on ultrasound images. International conference on Biomedical and Pharmaceutical Engineering, 88–92, December 2006
9. Shan, J., Wang, Y., Cheng, H.D.: A novel automatic seed point selection algorithm for breast ultrasound images. International Conference on Pattern Recognition, 1–4, 2008
10. Yu, Y.J., Acton, S.T.: Speckle reducing anisotropic diffusion. IEEE Trans. Image Process. **11**, 1260–1270 (2002)
11. Haralick, R.M.: Statistical and structural approaches to texture. IEEE Proc. **67**(5), 786–804 (1979)
12. Shi, X., Cheng, H.D., Hua, L., Jua, W., Tian, J.: Detection and classification of masses in breast ultrasound images. Digital Signal Proc. **20**, 824–836 (2010)
13. Shan, J., Wang, Y., Cheng, H.D.: Completely automatic segmentation for breast ultrasound using multiple domain features. Proceedings of 17th International Conference on Image Processing, 1713–1715, 2010
14. Ultrasound Cases Information. http://www.ultrasoundcases.info/
15. Ultrasound Images. http://www.ultrasound-images.com/breast.htm

EEA-LEACH—Trustworthy-Enhanced Algorithm for Energy Optimization in Wireless Sensor Networks

Umang, M.N. Hoda and Geetanjali Singh

Abstract Due to multifunctional nature of sensor nodes, the energy consumption among nodes increases gradually. Based on this issue, in this attempt, a new enhanced algorithm EEA-LEACH is proposed which includes basic concepts of energy-efficient LEACH algorithm and trustworthy cluster-based routing algorithms for large and small in network scenario. Proposed algorithm is an optimal solution and satisfies the encountered constraints related to energy conservation and security. It also balances the overload on a cluster head in the network. Analytical and simulation results show that suggested protocol can minimize the energy consumption among sensor nodes and increase the performance of network during data. In this attempt, MATLAB simulator is used here to judge the performance of the proposed algorithm EEA-LEACH and its comparative simulation result analysis with existing LEACH protocol.

Keywords Routing protocols · LEACH protocol · Energy consumption · Security · MATLAB simulator

1 Introduction

A sensor network comprises of several small devices called sensor nodes that are distributed independently in any diverse location to communicate wirelessly and examine physical or environmental conditions such as weather parameters,

Umang (✉)
Institute of Technology & Science, Mohan Nagar, Ghaziabad, India
e-mail: singh.umang@rediffmail.com

M.N. Hoda
Bharati Vidyapeeth Institute of Computer Applications, New Delhi, India
e-mail: mca@bvicam.in

Geetanjali Singh
M. Tech Student, Banasthali University, Jaipur, India
e-mail: geetanjalisingh829@gmail.com

S. Urooj and J. Virmani (eds.), *Sensors and Image Processing*, Advances in Intelligent Systems and Computing 651, https://doi.org/10.1007/978-981-10-6614-6_15

electromagnetic waves, and noise levels. The data gathered by these nodes are then transferred to the base station from where it is extracted by the user [1–5]. The purpose behind the invention of this technology is to embed numerous sensor nodes into the physical world to perform higher level identification and tasks. It has a wide range of applications in medicine departments, agricultural fields, military and environmental monitoring, and so on [6, 7]. Each node in a sensor network has a sensing material which is known as sensors, a transceiver, memory, a microcontroller, and a power source. The energy required for any sensor node to operate is very limited due to the use of batteries as the power source. Eventually, the battery runs out of energy and the node dies which may result in a network failure. Therefore, energy has a great significance in the deployment of a sensor network. To achieve the desired outcome from the network, it is very important to balance the energy consumption of sensor nodes. One way to accomplish this is to find a reliable routing method of data packets from the nodes to the base station. For this purpose, many routing protocols for sensor network applications have been suggested and discovered that are able to preserve the energy consumed by nodes and enhance the system performance [8].

The factors that consume most of the energy in a network:

- The communication in the network has much higher energy consumption than the manipulation and computation of data. Transmitting one bit of data may consume energy as much as performing a few hundred instructions.
- Depending on the requirements of different applications, sensing real data requires the consumption of energy.
- Data aggregation in a network also consumes energy.

The length of time during which the sensor network functions properly can be extended by adopting distinct optimization techniques. One of such techniques is the energy-enhancing-routing protocols which are successful in minimizing the energy expenditure during transfer of data [9–11]. Also, sensor nodes absorb a significant amount of energy even when they are not performing any task. The node components that are not in use can be switched off by applying some power management schemes. Based on the above factors, several approaches have to be appointed to reduce power consumption in a sensor network. Study of routing protocols has emerged as a prominent field in the sensor networks.

LEACH has been observed to be a very popular routing protocol that uses clustering as an optimization technique to enhance the network lifetime by reducing the use of energy. In this paper, Sect. 2 reviews literature survey. Then, Sect. 3 presents details of our algorithm. Therefore, Sect. 4 presents the performance of EEA-LEACH by simulations and compare it results with LEACH, and finally, Sect. 5 concludes the paper.

2 Literature Review

In wireless sensor network [12–14], routing protocols have mainly eight categories: (i) location-based routing (ii) data centric-based routing (iii) hierarchical routing (iv) mobility-based routing (v) multipath-based routing (vi) heterogeneity-based routing (vii) QoS-based routing and (viii) negotiation-based routing protocol. Xuxun Liu [15] has reviewed clustering routing protocol in wireless sensor networks in his work. In this paper, author surveyed clustering routing protocols along with their advantages and objectives for WSN. Further, comparative analysis along with cluster attribute is also discussed in detail.

In [16], author has proposed S-SPIN (Secure-SPIN) protocol by incorporating MAC Scheme for ensuring the guarantee and correctness of messages. This scheme is secure in case of existential forgery attacks only. Christina [17] has proposed an algorithm WBAODV based on dynamic key cryptography. In this approach, route is decided by three factors the speed of nodes, the power level of battery and the Bandwidth. And three levels are used for communication: (i) sensor to cluster, (ii) cluster to head and (iii) head to sink. However, this protocol may not be suitable for large network scenario.

In this paper [18], secure source-based loose synchronization (SOBAS) protocol is proposed which uses en route filtering mechanism, where the malicious data is filtered from the network. In this approach, nodes use their local time values as a one-time dynamic key to encrypt each message. Due to incorporation of loose synchronization, SOBAS reduces number of control messages which reduces energy consumption as well as reducing the number of malicious nodes to make the network secure. Reza et al. [19] propose a secure clustering scheme along with a deterministic pairwise key management scheme based on public key cryptography. This mechanism guarantees that any two nodes residing in same cluster can establish routing path directly by generating pairwise key without disclosing any information to other nodes. However, overhead may increase in case of broadcast from cluster gateways and frequent disconnected networks.

Abo-Zahhad et al. [20] has focused on energy hole problem. Suggested problem is overcome by proposing novel algorithm and mobile sink-based adaptive immune energy-efficient clustering protocol (MSIEEP) by using adaptive immune algorithm (AIA) into an account. This approach provides optimum ways to find cluster heads which improves lifetime and stability of network.

R. Velmani and Kaarthick [21] has proposed Velocity Energy-efficient and Link-aware Cluster Tree (VELCT) scheme for data collection in wireless sensor networks. This scheme creates data collection tree (DCT) based on location of cluster head. Data collection node simply collects data from cluster head and delivers it to the sink. Suggested scheme minimizes energy exploitation, reduces end-to-end delay and traffic in cluster head, and also avoids frequent cluster formation.

Taghikhaki, Z. has proposed reliable and energy-efficient chain cluster-based routing named as REC+ which provides maximum reliability in multihop network by defining proper shape and size of cluster and proper place to cluster head by considering parameters' energy efficiency, transmission reliability, intra-cluster delay, and cluster head selection into an account.

3 Improvement of LEACH

3.1 *Preliminary Concept*

(i) Entropy: assume that X a random variable which takes on a finite set of values $x1, x2, x3\ldots xn$ with probability $P(x = xj) = pi$ where $0 \leq p \leq 1$ and $1 \leq i \leq n$ & where $\sum pi = 1$
Entropy X is an amount of information provided by an observation of X
X can be defined as
$H(X) = -\sum_{i}^{n} \text{pilog}\left(\frac{1}{\text{pi}}\right)$

(ii) **Cross-entropy:** this Monte Carlo approach has two phases (i) randomly generating date (ii) updating the parameter of random mechanism
$D = \text{Eu}[H(X)] = \int H(X)f(x;\ i)\text{dx}$

$$D = \frac{1}{N} = \sum_{i=1}^{N} H(Xi)f(Xi,\ u)/g(Xi) - \frac{1}{N}\sum_{n=1}^{N}\left[yn\log yn' + (1 - yn)\log\left(1 - yn'\right)\right]$$

Cross-entropy between sender (s) and receiver (r) over same set of events calculates average number of bits = $H(s, r) = Es[-\log r] = h(s) + D_{KL}(s\|r)$
where $H(S) = \sum_{i=1}^{n} \log\left(\frac{1}{si}\right)$

(iii) **Trust:** it is represented as probability associated to binary relation between primary node and cluster head CH.

3.2 *Problem Statement*

Low-energy adaptive clustering hierarchy (LEACH) is a routing protocol which was discovered as the most prevalent energy-efficient TDMA-based protocol for sensor networks. It uses clustering as a foundation to carry out its tasks. The goal of this protocol is to lower the power consumption required for maintaining the clusters to expand the duration of time in which the network functions [22–27].

LEACH protocol is faced with many difficulties. One of them is the overload of work on a cluster head. In this protocol, any node can be selected as a CH in a round. The work of a CH is to collect data from all the nodes within its cluster, compress that data, and forward it to the base station. Even after the compression and aggregation, the data quantity is quite large. The communication in sending this data to the base station requires a lot of energy. This energy consumption is higher than the data manipulation and compression. As the energy level of a CH decreases drastically after performing these functions, it dies quickly. While conducting a sensor network application, if a CH fails rapidly and before time, then there will be no profit resulting in the failure of the application [26, 28].

3.3 *EEA-LEACH Design*

The EEA-LEACH scheme consists of set-up phase and steady-state phase. In set-up phase, selection of cluster head and cluster formation is initiated, and selection criteria is based on important parameter, i.e., transmission range, distance calculation, node degree calculation, and primary node selection criteria and energy conservation. Each cluster head appoints a primary node from all the nodes in its cluster. Primary node will remain active during this phase. Further, in steady phase, primary node is having onus to share the information their nearby neighbour nodes to cluster head and transfer the information to the destination node (Figs. 1, 2 and 3).

The above figure explains the procedure of collecting data and then forwarding it for further analysis.

1. The sensor nodes monitor the environment and collect the data via sensors.
2. They forward this data to the primary node.
3. The primary node then gathers all the data from the nodes of the cluster to perform aggregation and compression. This data is then forwarded to the cluster head.
4. The cluster head collects the data from the primary node and sends it to the base station (Fig. 4a, b).

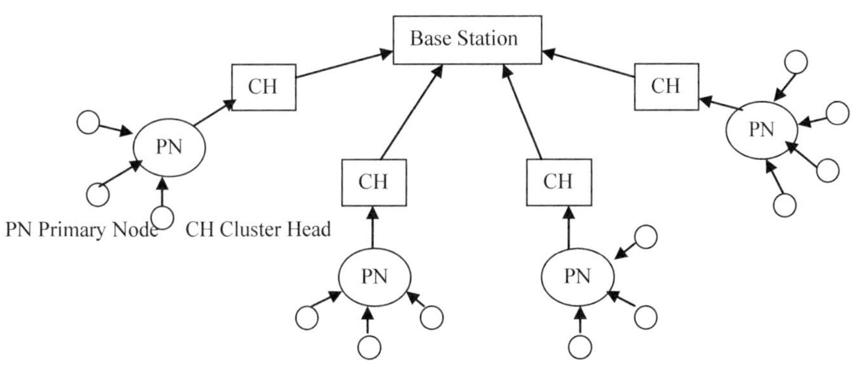

Fig. 1 Architecture of EEA-LEACH

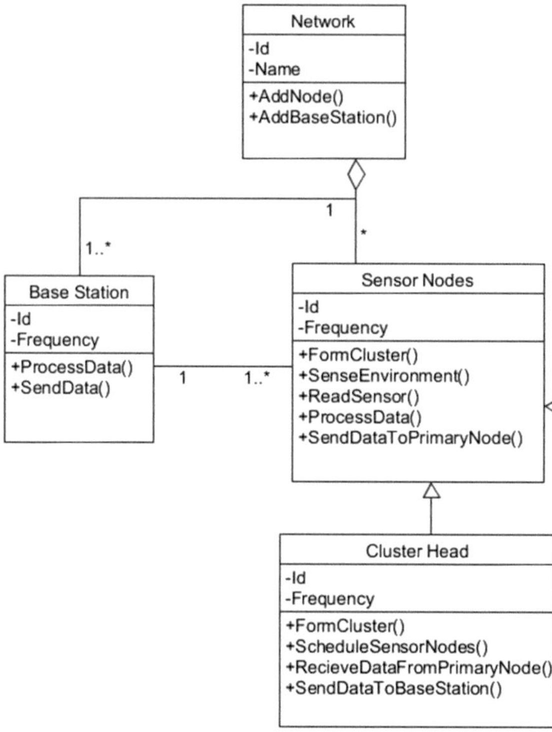

Fig. 2 Class diagram of EEA-LEACH

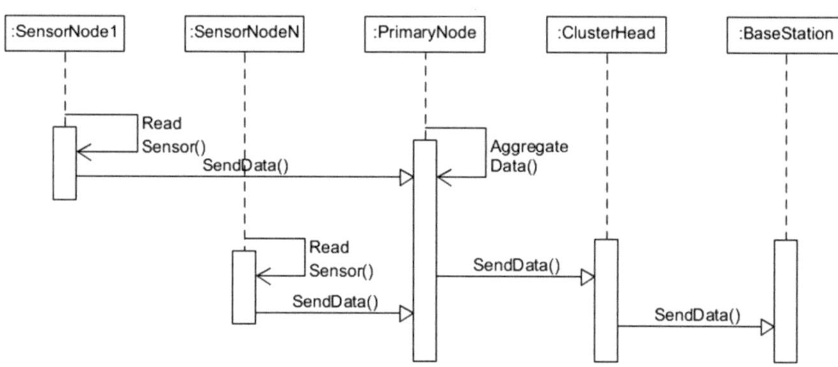

Fig. 3 Sequence diagram for use case of data gathering and transmission

(a)

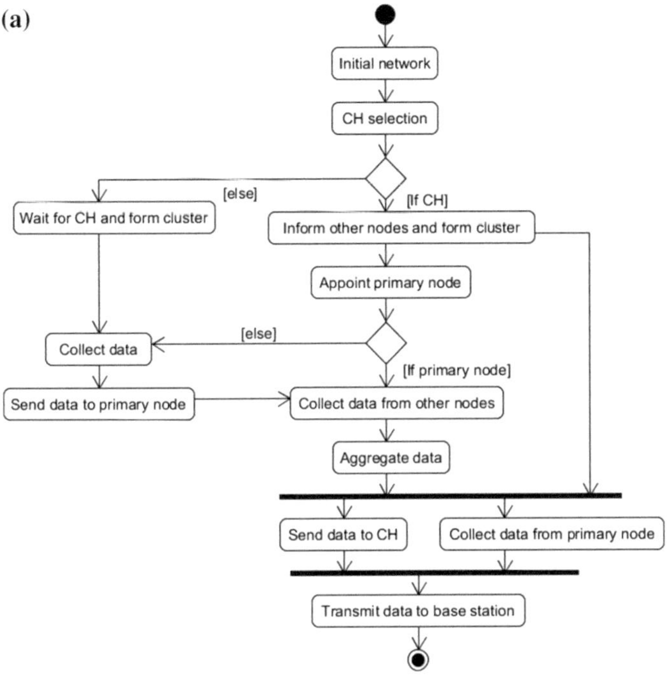

(b)

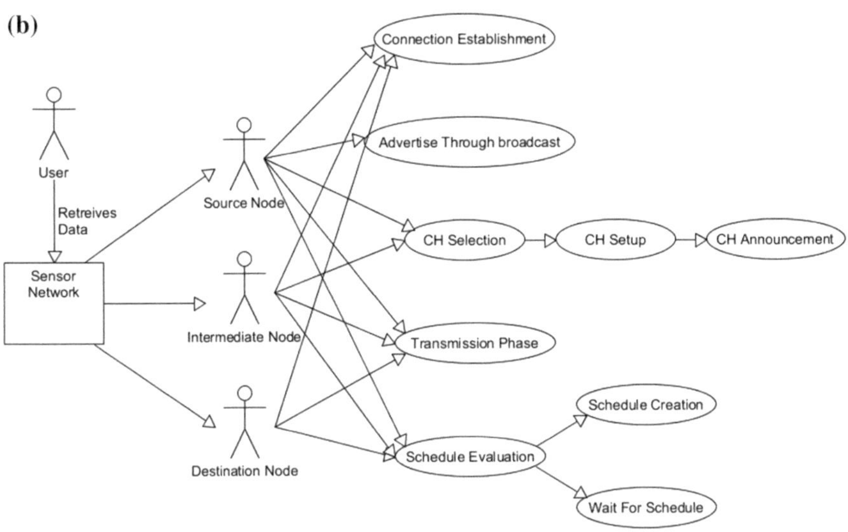

Fig. 4 **a** Activity diagram showing the processes of EEA-LEACH. **b** Use case diagram depicting the working of nodes in EEA-LEACH

3.4 Terminology Used

3.4.1 Calculate Transmission Range and Distance of Cluster Head with Primary Node

We have used Euclidean distance formula for distance calculation between selected cluster head CH and primary node lying in the network.

$p1$, $q1$ are the p, q coordinates for one side of a line, and $p2$, $q2$ are on the other side, and distance d is the distance between two points.

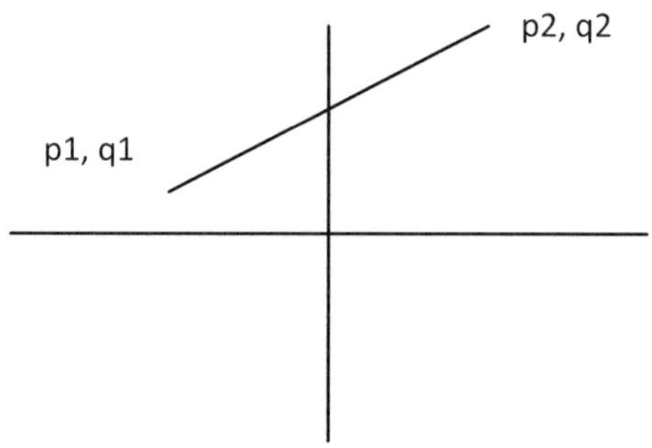

$$d(p,\ q) = \sqrt{(p1 - q1)^2 + (p2 - q2)^2 + \ \cdots \ + (pn - qn)^2} = \sqrt{\sum_{i=1}^{n} (pi - qi)^2}$$

Degree of node is the number of connected nodes. Adjacent nodes are having *depth* = 1 and nodes which are adjacent to adjacent node will have *depth* = 2.

3.5 Energy-Efficient Advanced LEACH Algorithm (EEA-LEACH)

To deal with such difficulties, trustworthy energy-efficient advanced LEACH algorithm (EEA-LEACH) balances the workload on a cluster head. The idea behind the improved algorithm is to introduce a primary node in a cluster whose job would be to share the work load of a cluster head. The method of selecting a cluster head and forming clusters is as discussed in Algorithm 1. Here, after the clusters are formed and CHs are elected, the CH of each cluster would appoint a node of its cluster as a primary node. The selection of primary node is done by following two criteria:

(i) All nodes and selected cluster head must be in same transmission range.
(ii) Node having minimum distance and max in and out degree in cluster head.
(iii) Calculate beacon response time of each primary node as trustworthy node
(iv) The residue energy of the node.

Each cluster head appoints a primary node from all the nodes in its cluster. The nearest node with maximum energy and minimum distance is elected as primary node. Each of encountered nodes will be enlisted into active nodes. The primary node broadcasts a message of being primary node to all the other nodes in the cluster and builds a TDMA schedule for receiving and transmitting data.

The work of a primary node will be to gather the data from all the nodes of the cluster. Then, aggregate this data and send it to the cluster head. Then, the CH will transmit that data to the base station. In this way, the work load will be divided between the primary node and CH. In every round, this process will be repeated, i.e., reelection of CH and primary node [28].

New technologies replace existing technologies when there are economic advantages. In the new improved algorithm, lower overhead will also provide cost advantages. Therefore, this algorithm can replace the older one and can be used in various sensor network applications. It can be used in medical departments to monitor diseases as well as provide tactical monitoring of enemy troops in military. It can also be used in industrial and commercial departments and other applications of sensor networks. Figure 1 shows the architecture of the improved algorithm.

This protocol is separated into rounds, where each round is composed of two distinct phases:

Set-up Phase
In this phase, any node is elected as a cluster head randomly for a round. Every node in a network has the liberty to decide on its own whether it wants to be a CH or not. After the selection of CH (as per Algorithm 1), it informs the neighbouring nodes to form a cluster by transmitting an advertisement packet. The other nodes in the network receive the advertisement packet. By using CSMA concept, these nodes send a join packet to the CH informing it to join its cluster. In this way, a cluster is formed by incorporating features of TDMA schedule during which the nodes will communicate with the CH. Each node has a different time slot to send data packets. This schedule is broadcasted by the CH to all the nodes in the respective cluster.

Algorithm for Cluster head Selection:

1. Scan all nodes present in the network with in Tr.
2. Broadcast beacon message
3. If node with Tr & Tp will send reply with in time interval T.
4. if True then
   a. Calculate Node in & out degree
   b. Calculate Transmission power
   c. Calculate Node Coverability

```
        If Ok
          Initialize nocn=0;
           Cluster=1;
          If  (nocn<=50) then              // no of nodes in cluster
        Then cluster= nocn+1
        End If
    Else
         Send message for time interval T
            Check If response received
            Call Step 4
            Else
            Session time expires
            Call Step 2
             End If
            End If
```

Steady Phase

Data transmission begins in this phase. The nodes of a cluster gather the sensory data from the real world and forward their data to the CH in their scheduled time slot. The data transfer requires minimum amount of energy. The CH aggregates the data after receiving it from all the nodes. The compression and aggregation of data is performed to remove any redundancy that might occur. This data is then forwarded to the BS.

4 Simulation

In this paper, MATLAB is used to conduct the simulation of LEACH protocol and the new algorithm EEA-LEACH. The objective of the proposed algorithm is to overcome the difficulties and problems faced by LEACH protocol. This is achieved by balancing the overload of work on CH and decreasing the total consumption of energy by a sensor network. The lifetime of any network increases when it is said to be energy efficient. The network lifetime means the duration of time within which a network performs properly. When the last node in a network dies, the network dies. To measure the performance of the algorithm, consumption of energy by nodes is taken as an indicator. Lower the energy consumption, higher the lifetime of network [29, 30].

4.1 Parameters Taken for Simulation

1. Distribution of sensor nodes at any location or region is random.
2. The nodes have similar properties and are homogeneous in nature.
3. They have limited amount of energy, and their position on the field is fixed.
4. The location of the base station is fixed at the region.
5. The nodes are able to communicate with the base station independently (Table 1).

4.2 Simulation Results

100 nodes are deployed randomly within a region of 100 * 100 m. The base station is located at a fixed point at (10, 50). Figure 5 shows the node distribution of a sensor network. It is clearly visible that the nodes are deployed in a random fashion.

Table 1 Parameters used for simulation

Parameters	Values
Area	100 * 100
Number of nodes	100
Initial energy	0.5 J
BS location	(10, 50)
CH probability	0.1
Packet size	4000 bits
Eelec	50 nJ/bit
Efs	10 pJ/bit/m^2
Emp	0.0013 pJ/bit/m^4
Eda	5 nJ/bit

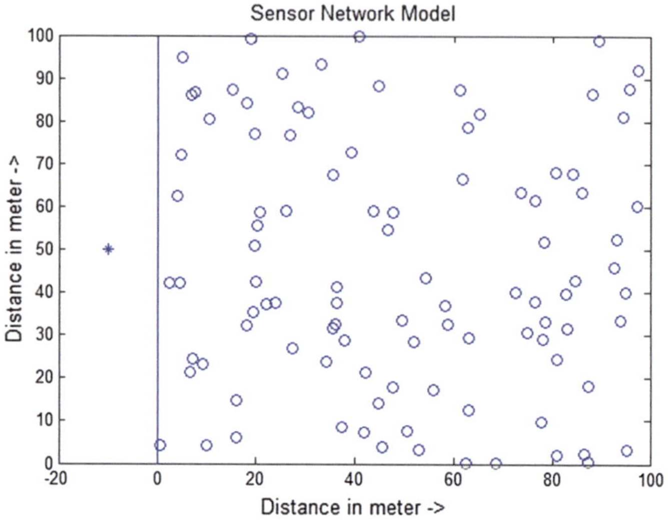

Fig. 5 Randomly distributed nodes

The Total Remaining Energy in a Network

The energy consumed by the network is shown in the following figures. It can be seen in Fig. 6 that EEA-LEACH algorithm performs better as the rate of its energy depletion is lower than that of LEACH protocol. The energy consumption of the network is reduced by setting the primary nodes reasonably. By the introduction of primary nodes, the overload on the cluster head is decreased. This helps in balancing the energy usage of the nodes and enhances performance of the network. In Fig. 6, the numbers of nodes taken are 100, and from the analysis of the figure, it is proved that the network dies early in LEACH protocol as its total network remaining energy depletes faster than EEA-LEACH.

The above figure shows the comparison between the energy consumption in a network between LEACH and EEA-LEACH when there are 100 nodes. When the numbers of nodes are decreased to 50, it is noticed that the energy level of the network also decreases, although the energy diminishes at the same number of rounds. Figure 7 shows the energy consumption of the network when the numbers of nodes are reduced to 50.

Energy Depletion of Cluster Heads

Figures 8 and 9 show the energy depletion of total cluster heads in a network as the number of rounds increases during simulation. In case of LEACH protocol, the energy on cluster heads is fully consumed after round 18 and they die. This is due to the overload of work on the CH that its energy is compromised.

Figure 9 shows the cluster heads in EEA-LEACH algorithm. The difference between both the figures is minor but noticeable. In case of EEA-LEACH, the CHs are able to sustain their lifetime as their work load is decreased by sharing it with

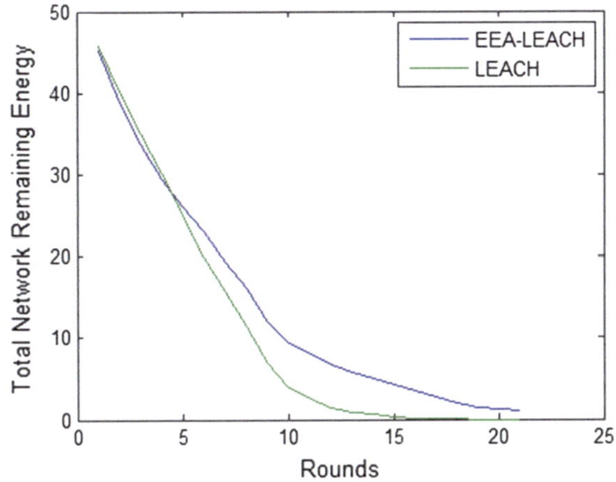

Fig. 6 Total remaining energy for 100 nodes

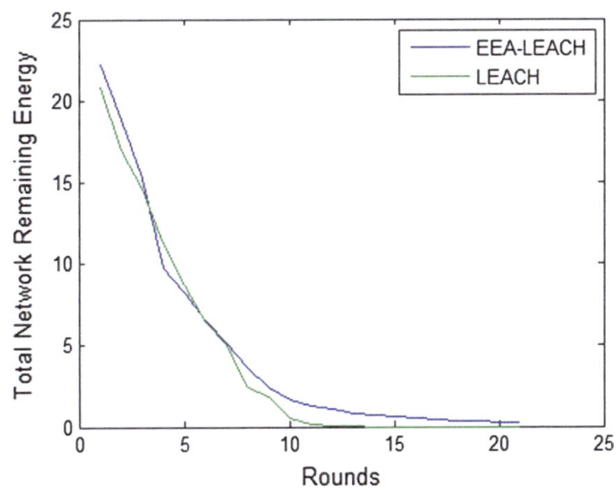

Fig. 7 Total remaining energy for 50 nodes

the primary node. The energy consumption by a CH is hence reduced and therefore it survives for a longer time.

The Remaining Number of Nodes

The number of nodes taken in the network is 100. When the energy of a node is completely utilized, it is unable to perform its tasks and is declared as dead. When all the nodes in a network die, the network no longer functions. If the nodes die before the functionality of a network is completed then the application in which the network was deployed fails. Therefore, the durability of nodes plays an important feature to judge the performance of a network. The numbers of nodes which are able to retain their energy after the simulation is known as the remaining nodes in a network. Figure 10 shows the remaining number of nodes in LEACH protocol which is found out to be only 1.

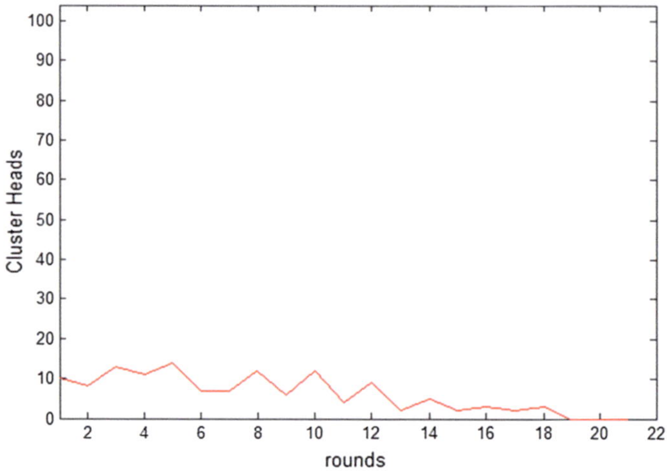

Fig. 8 Energy depletion of cluster heads in LEACH

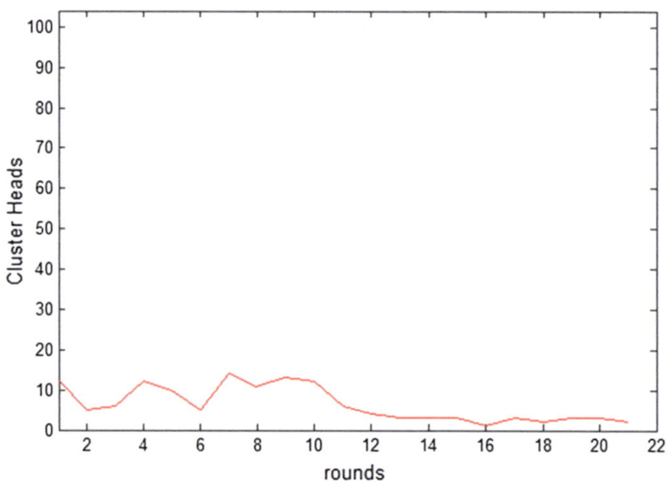

Fig. 9 Energy depletion of cluster heads in EEA-LEACH

Figure 11 shows the remaining number of nodes in EEA-LEACH algorithm which is more than that of LEACH. This implies that in EEA-LEACH, nodes are able to retain their energy for a longer span of time which results in better performance of the network as it increases its lifetime.

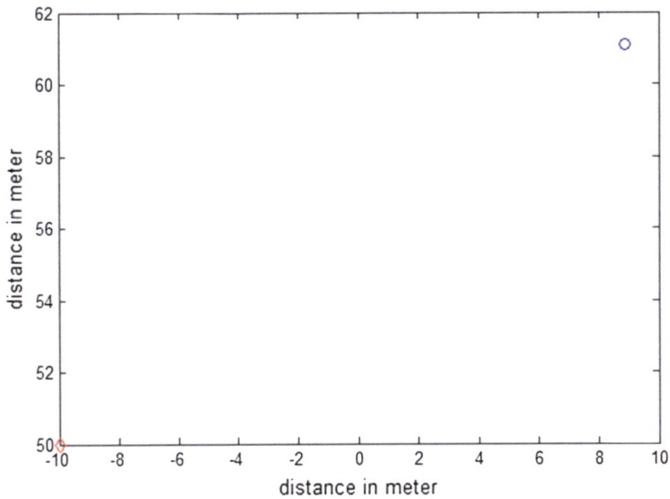

Fig. 10 Remaining number of nodes in LEACH

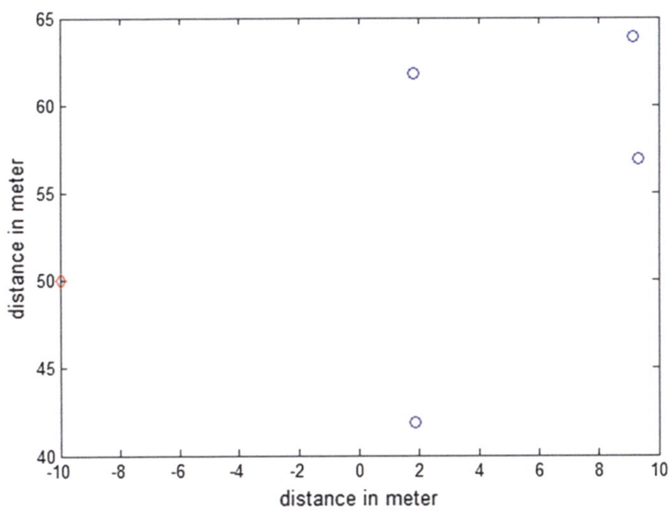

Fig. 11 Remaining number of nodes in EEA-LEACH

5 Conclusion

The growing popularity of sensor networks in various applications such as military and medicine fields has increased the research of routing protocols in WSN. The objective behind designing such protocols is to prolong the functionality of nodes until the desired result is obtained for a particular application. If the nodes fail before completing their task, then it would lead to poor performance of the whole network system. The durability and lifetime of a node depends upon the energy consumed by it during data transmission, and due to the limited energy resources, the lifetime of nodes and the network is compromised. Therefore, the routing protocols should be energy efficient so that they can increase the scalability of the network and also prolong its life span.

In this paper, EEA-LEACH algorithm is proposed to decrease the overload on a cluster head and prolong the sensor network lifetime. LEACH protocol is used as the foundation of this algorithm. The objective is achieved by introducing a primary node and dividing the work of CH between them. Due to this, the energy on CH is retained for a longer span of time. It would avoid the reelection of CH frequently. Battery power on a CH would also increase as its work overload is decreased. The new EEA-LEACH algorithm is judged by performing simulation in MATLAB environment. The simulation results point out the energy efficiency of the new algorithm and prove that its functionality is better than that of traditional LEACH protocol.

References

1. Akyildiz, I.F., et al.: Wireless sensor networks: a survey. Comput. Netw. **38**, 393–422 (2002)
2. Sohrabi, K., et al.: Protocols for self-organization of a wireless sensor network. IEEE Pers. Commun. **7**(5), 16–27 (2000)
3. Min, R. et al.: Low power wireless sensor networks. In the Proceedings of International Conference on VLSI Design, Bangalore, India, January 2001
4. LEWIS, F.L.: Wireless Sensor Networks, Associate Director for Research, The University of Texas at Arlington
5. Yick, J., Mukherjee, B., Ghosal, D.: Wireless sensor network survey. Comput. Netw. **52**(12), 2292–2330 (2008)
6. Keshtgary, M., Deljoo, A.: An efficient wireless sensor network for precision agriculture. Can. J. Multimedia Wireless Netw. **3**(1), 1–5 (2012)
7. Yuce, M.R., Ng, P.C., Lee, C.K., Khan, J.Y., Liu, W.: A wireless medical monitoring over a heterogeneous sensor network, Engineering in Medicine and Biology Society, 2007. EMBS 2007. 29th Annual International Conference of the IEEE, 5894–5898 (2007)
8. Mundada, M.R., Kiran, S., Khobanna, S., Varsha, R.N., George, S.A.: A study on energy efficient routing protocols in wireless sensor networks. Int. J. Distrib. Parallel Syst. **3**(3), 311–330 (2012)
9. Anastasi, G., Conti, M., Di Francesco, M., Passarella, A.: Energy conservation in wireless sensor networks: a survey, Department of Information Engineering, University of Pisa, Italy
10. Sendra, S., Lloret, J., García, M., Toledo, J.F.: Power saving and energy optimization techniques for wireless sensor networks. J. Commun. **6**(6), 439–459 (2011)
11. Singh, S.K., Singh, M.P., Singh, D.K.: Routing protocols in wireless sensor networks—a survey. Int. J. Comput. Sci. Eng. Surv. (IJCSES) **1**(2), 63–83 (November 2010)
12. Akkaya, K., Younis, M.: A survey on routing protocols for wireless sensor networks. Ad Hoc Netw. **3**(2005), 325–349 (2003)
13. Patil, M.: A survey on routing protocols in wireless sensor networks 18th IEEE International Conference on Networks (ICON), 86–91, 978-1-4673-4521-7. doi:10.1109/ICON.2012. 6506539 2012, IEEE

14. Singh, S.P., Sharma, S.C.: A survey on cluster based routing protocols in wireless sensor networks. International Conference on Advanced Computing Technologies and Applications (ICACTA-2015), proceedings Computer Science **45**(2015), 687–695, 1877-0509 (2015)
15. Liu, X.: A survey on clustering routing protocols in wireless sensor networks. Sensors 11113–11153, ISSN 1424-8220, (2012). doi:10.3390/s120811113
16. Tang, L.: S-SPIN: a provably secure routing protocol for wireless sensor networks. International Conference on Communication Software and Networks. ICCSN'09, 620–624, 978-0-7695-3522-7 (2009)
17. Christina, D.P.S.E.: Energy efficient secure routing in wireless sensor networks. International Conference on Emerging Trends in Electrical and Computer Technology (ICETECT), 982–986 ISBN 978-1-4244-7923-8, (2011). doi:10.1109/ICETECT.2011.5760262
18. Uluagac, S.A.: Secure Source-BAsed Loose Synchronization (SOBAS) for wireless sensor networks. Parallel Distrib. Syst., IEEE Trans. **24**(4), 803–813, ISSN: 1045-9219 (2011)
19. Azarderskhsh, R., Reyhani-Masoleh, A.: Secure clustering and symmetric key establishment in heterogeneous wireless sensor networks. EURASIP J. Wireless Commun. Netw. (2011). doi:10.1155/2011/893592
20. Abo-Zahhad, M., Ahmed, S.M., Sabor, N., Sasaki, S.: Mobile sink-based adaptive immune energy-efficient clustering protocol for improving the lifetime and stability period of wireless sensor networks. IEEE Sens. J. **15**(8), 4576–4585 (August 2015)
21. Velmani, R., Kaarthick, B.: An efficient cluster-tree based data collection scheme for large mobile wireless sensor networks. IEEE Sens. J. **15**(4), 2377–2390 (April 2015)
22. Heinzelman, W.R., Chandrakasan, A., Balakrishnan, H.: Energy-efficient communication protocol for wireless microsensor networks. Proceedings of the 33rd Hawaii International Conference on System Sciences—2000
23. Heinzelman, W.B., Chandrakasan, A.P., Balakrishnan, H.: An application specific protocol architecture for wireless microsensor networks. IEEE Trans. Wireless Commun. **1**(4), 660–670 (2002). doi:10.1109/TWC.2002.804190
24. Jaiswal, V.P.N., Dr. Garg, A.K.: An efficient protocol for reducing energy consumption in wireless sensor networks. Int. J. Eng. Res. Appl. (IJERA) ISSN: 2248–9622, **2**(3), 530–533 (May–June 2012)
25. Gnanambigai, J., Dr. Rengarajan, N., Anbukkarasi, K.: Leach and its descendant protocols: a survey. Int. J. Commun. Comput. Technol. **1**(3), 15–21 (2 September 2012)
26. Chunyao, F.U., Jiang, Z., Wei, W., Wei, A.: An energy balanced algorithm of LEACH protocol in WSN. Int. J. Comput. Sci. **10**(1), 354–359 (1 January 2013)
27. Braman, A., Umapathi, G.R.: A comparative study on advances in leach routing protocol for wireless sensor networks: a survey. Int. J. Adv. Res. Comput. Commun. Eng. **3**(2) (February 2014)
28. Rauthan, R.S., Mishra, S.: An improved approach in clustering algorithm for load balancing in wireless sensor networks. Int. J. Adv. Res. Comput. Eng. Technol. **1**(5) (July 2012)
29. Singh, G., Dr. Umang.: Optimization techniques for energy consumption in sensor networks-survey. Int. J. Comput. Appl. (0975–8887) Emerging Paradigms of Information and Communication Technologies and its Impact on Society (EPICTIS-2014)
30. Kaur, A., Saini, S.: Simulation of low energy adaptive clustering hierarchy protocol for wireless sensor network. Int. J. Adv. Res. Comput. Sci. Softw. Eng. **3**(7), 1316–1320 (July 2013)

Secure Cluster-Based Data Integrity Routing for Wireless Sensor Networks

Shashikala and C. Kavitha

Abstract The wireless sensor network (WSN) is a resource-limited device, which majorly looks into the security and the lifetime of the sensors. The protocol designed here aims at the prolonging the lifetime of the sensors and at the same time providing the security of the data sent from source to sink in the cluster. There are various cluster-based protocols are available, most of the protocols consider the cluster area as a circle, hexagon, and so on, which is regular in shape. The major drawback with regular shape clusters is isolated nodes are more in the network, and the nodes may be placed at the boundary in which case the node may send the data to more than one cluster head, the data is secured both inside the cluster and across the cluster in which case the energy consumptions are more. The proposed protocol shows that the clusters of irregular shapes, with ideal distance and securing the data only across the cluster, can improve the lifetime of the nodes in the networks.

Keywords WSN · Secured · Cluster · Routing protocol

1 Introduction

The sensors are used to sense the various information, such as heat, light, movement, vibration, location, touch, temperature, and so on, to achieve these things efficiently the sensors are distributed across the sensing region which may be of size as small as our palm or as big as a city. Spatially distributed sensor nodes in the environment form a self-configuring, self-organizing network. The nodes can communicate with the sink node using long path, which in turn takes more time, and bleach the security of the data and the network lifetime is also very small, and to overcome this the clusters are used in the network. The set of nodes is grouped to form the network, which is called a cluster. The cluster will have the cluster

Shashikala (✉)
BNMIT, Department of CS, Bangalore, India

C. Kavitha
BNMIT, Department of ISE, Bangalore, India

S. Urooj and J. Virmani (eds.), *Sensors and Image Processing*, Advances in Intelligent Systems and Computing 651, https://doi.org/10.1007/978-981-10-6614-6_16

members and the cluster head. The cluster members will collect the information and send to the cluster head and cluster head will, in turn, send the data to the sink. The cluster formation helps in many ways that it avoids the long-path security breach for the data and also prolongs the life of the network, network scalability, and network topology.

There are various methods available to form the cluster, that is, selecting the cluster head and the cluster members. The cluster members are selected either based on the location or the radio range of the sensor nodes. The cluster head is elected based on the residual energy or the ideal distance between the nodes of the cluster members in regular geometric shape. The cluster area is considered to be a regular shape which may lead to many isolated nodes in the network.

The cluster is formed in three different ways as follows: *block based cluster*, in this, the clusters are of predefined shapes and each cluster knows its area of coverage. In *Grid based cluster*, the sensing region is dived as a grid; in this, also the sensors know the sensing area, but in these clusters disadvantage is that the clusters are formed statically. In *chain based cluster*, the clusters are formed dynamically, and the advantage is that the size of the cluster can be varied based on the data accuracy required or a number of sensors [1]. The examples of different type of cluster are given in Table 1.

To ensure the data integrity of the network, the data transmitted between the members and head of the cluster is encrypted. Then each node will encrypt the data and transmit to the cluster head, and cluster head can, in turn, transmit the data as it is to sink or it may again check correctness of the data and then transmit, which consumes more energy for each node. But the cluster head has the knowledge about the location of the cluster members and the available energy of the node, if any intruder sends the data, then it can be easily identified by the cluster head and drop that node from the network [1, 2, 3, 4].

This paper proposes a protocol which elects the cluster member and head from and irregular geometric shape to avoid the isolated nodes in the network and to prolong the lifetime of the node in the network. The cluster head is elected based on the residual energy and the finest distance between the CH and all other nodes in the network, here the cluster node will aggregate the information and just forwards it toward the head, and the cluster head will generate the hash value and then sends to sink. The simple illusion is as shown in Fig. 1.

The paper is organized in the following manner: Sect. 1 is an introduction, Sect. 2 is related work and the major problems in the sensor network, Sect. 3 is

Table 1 The routing protocol with different types of clusters

Block based cluster	Grid based cluster	A chain based cluster
LEACH	PANEL	PEGASIS
HEED	GAF	CCS
UCS	TTDD	TSC
EECS	SLOC	

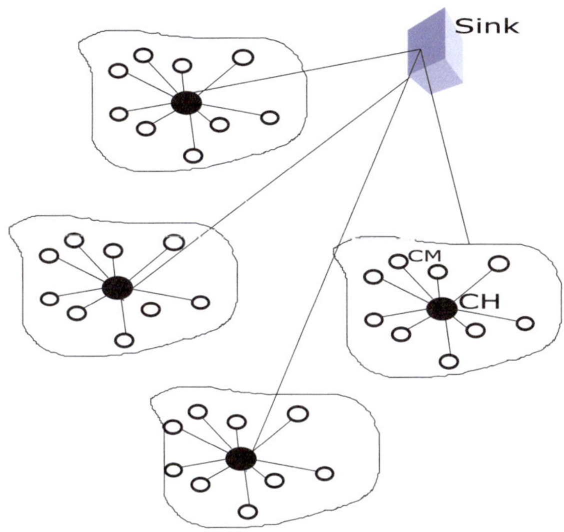

Fig. 1 Clusters showing the communication between Cluster Head (CH), Cluster Member (CM) and Sink

execution of the work done in detail, and Sect. 4 deals with the execution analysis, which is followed by a conclusion.

2 Related Work

The data is sent between the cluster head and the sink, the data transmission uses the routing algorithm. The design of the routing algorithm as various design challenges such as node deployment, energy consumption, nature of node, coverage, scalability, quality of service, and application which is using the sensors [1] and the security either on the data or for the routing path.

Node Deployment: The node deployment depends on the application in use. The nodes can be placed randomly or in a precise way. The random node deployment is useful in harsh environment, but it does not cover the entire sensing region. Where as, in the other case, the nodes will cover the entire region, and this method cannot be used in the harsh environment.

Nature of Node: This again depends on the application in use. The application may demand the homogeneous or heterogeneous nodes. The homogeneous nodes all will have the same capabilities, but in heterogeneous, the nodes will have different capacities.

Energy Consumption: As the wireless sensor nodes are resource constraints, the energy in the sensors is limited. Thus, we need to design the routing algorithm, which prolongs the lifetime of the sensors.

Coverage: The sensing region is usually large and we need to scatter the sensors throughout the region densely to cover the entire region and avoid the isolated node in the region.

Scalability: The sensors are resource-limited which may lead to dead nodes in the network, which in turn lessens the number of nodes in the network. Then the sensors need to form a new topology based on the available node or to get more accurate results; the end user may add more number of sensors in the region, then also the network need to form a new topology to transmit the data. The way the network formed by these sensors should be efficient.

Security: The sensor network need to secure both data and the route via in which it sends the information [5, 1].

The cluster-based energy efficient routing in WSN is energy efficient as the cluster head elected based on the residual energy and the optimal distance from all nodes in the cluster and it uses a homogeneous node for communication. Compared to LEACH and HEED, this protocol saves more energy. However, this protocol gives downward performance for dynamic formation of clusters and for the security of the data [2].

The cluster is elected based on two parameters: first on the residual energy and then from the communication distance between the sensor nodes using RSSI. In this, the base station collects the RSSI value of sensor node and finds the average value based on it, and then it divides the network into primary tier and secondary tier. From the secondary tier, it elects three nodes as cluster head. Cluster head to communicate between cluster node and cluster head, this is identified as T1; to forward the data between primary tiers is identified as G1; and the third one to communicate between primary tier and secondary tier and this is known as G2. The drawback is, as it is using three nodes, to perform the cluster head operation in which the energy may deplete very fast in each round [2].

The distributed cluster-based protocol uses ROL/NDC to form a cluster. These protocol groups the nodes in the cluster and build the routing path based on the linear metrics to the sensor network. This protocol best suited for the application, which demands for the quality of services such as data aggregation and image transmission or detection [4].

Efficient, intelligent energy routing protocol in WSN based on the reinforcement learning techniques. Initially, it forms the cluster based on the FTIEE. It is a hierarchical based protocol and a connected graph is used to establish a network. The data transmission is by using the Q-value parameters of reinforcement learning technique. The protocol uses the machine learning process, which can be improved by using the fuzzy logic or Bayesian network [6].

To form a cluster in WSN, it requires additional energy consumption, but the Regional Energy Aware Clustering with Isolated Nodes (REAC-IN) uses a different method. The Cluster head selected is based on the weight. The weight calculated is based on the residual energy and the regional average energy of all sensors in each cluster [7].

The major problem in the cluster-based wireless sensor network is, residual energy, finest distance between the nodes and the security bleach.

Residual energy: Cluster head in WSN is elected based on the residual energy of the cluster head and cluster members. The nodes will periodically send the available energy in it to the neighbor nodes. The cluster head is the node that has the highest residual energy. The drawback of this method is all nodes may not have direct path due to this some nodes may become isolated nodes or when multi-hop routing is used, the intermediate nodes may lose their energy more faster. The other problem is that the intermediate nodes may compromise with the intruder.

The finest distance between the nodes: Rather selecting only the residual energy to decide about the cluster head, the distance between the nodes are also considered to elect the cluster head and the advantage is the isolated nodes are lessened when compared with the residual energy method and it avoids the energy lose in the intermediate nodes. This method avoids the isolated nodes, but the intruder or attacker can still exist in the networks.

Security bleaches: By securing the routing path or by securing the data what we transmit in it, we can achieve security. By applying the symmetric or asymmetric cryptography algorithm, we can achieve data security.

3 Secure Cluster-Based Data Integrity Routing for Wireless Sensor Networks

This section presents the details of the Secure Cluster-based Data Integrity Routing (SCDIR) algorithm. In this protocol, the cluster formation and communication are achieved in a secure way in three different phases. In the first phase, the residual energy is shared among the members of the clusters. In the second phase, the finest distance between the members of the cluster is taken care. Finally, to secure the data the cryptography algorithm is used.

We assume that initially, the nodes will exchange the distance and the initial energy in the node with all its neighbor nodes. All nodes will store this information in the routing table. Then cluster is formed with those nodes, which is located within its radio range of the node and has highest residual energy.

A. *Residual energy with ideal distance in the network*

Periodically, the residual energy or the available energy in the network is calculated. The nodes in each cluster will send the residual energy to its cluster member. The energy consumption in each sensor node calculates based on the Eq. 1 [5].

The energy consumption of transmitting data is

$$E_{TX}(k,d) = E_{elec} * K + \varepsilon_{amp} * K * d^2 \, for \, d > 1 \quad (1)$$

And the energy consumption to receive data is

$$E_{RX}(k,d) = E_{elec} * K \tag{2}$$

where K is the data transmitted in bits, d is the distance between the two sensors, E_{elec} is the energy consumption to carry out data transmission in terms of nJ /bit, and $E_{elec,}$, is the energy consumption constant used to expand radio coverage in terms of nJ/(bit * m^2).

From Eqs. 1 and 2 the total energy consumed is from the node is calculated by using Eq. 3.

$$= \sum E_{TX} + \sum E_{RX} \tag{3}$$

The residual energy (R_e) given by the Eq. 4.

$$\text{The } R_e = \text{Initial energy} - \text{Total energy consumed} \tag{4}$$

The above R_e gives the residual energy of the whole networks, but it is required to calculate the residual energy of each cluster. The residual energy in Eq. 4 is for the infinite time bound. Whereas in the cluster, the sensed data transmitted is over the finite distance, thus the time required to transmit the data is finite and depends on the distance, the number of intermediate nodes and the threshold value considered.

$$T_t = \frac{D_t}{R_t} \tag{5}$$

$$T_r = \frac{D_r}{R_t} \tag{6}$$

where T_t and T_r are transmit and receive times respectively, D_t and D_r are the static data bits transmitted and received within the cluster. R_t is the transmit or receive rate in the network.

As data transmission is static, therefore

$$D_t = D_r = D \tag{7}$$

In each cluster, the maximum amount of data sent are given by

$$N = \frac{1}{2} * R_t \tag{8}$$

The residual energy of the cluster after each second is

$$CR_e = \sum R_e - N \tag{9}$$

$$if \begin{cases} CR_e < Threshold, & then\ cluster\ can\prime t\ transmit\ any\ more\ data\ and\ cluster\ is\ dead \\ CR_e \geq Threshold, & then\ cluster\ can\ send\ the\ data \end{cases}$$

These residual energies calculated in the Eq. 9 gives the residual energy of each cluster.

B. *The ideal distance between the nodes*

The clusters are, usually, almost the same shape to communicate with the members. In this protocol, the clusters are of irregular shape. The advantage of using the irregular shape of the cluster is to lessen the number of isolated nodes in the network. Based on the Euclidean n-space method, the distance between the member nodes of cluster is computed.

The Euclidean distance between the two nodes, P and Q, is the length of the line segment connecting the PQ. The Euclidean n-space is given in the Eq. 10.

$$D = d(P,Q) = d(Q,P) = \sqrt{(\mathrm{q1}-\mathrm{p1})^2 + (\mathrm{q2}-\mathrm{p2})^2 + \cdots + (\mathrm{q2}-\mathrm{p2})^2} \quad (10)$$

$$= \sqrt{\sum_{i=1}^{n} (qi - pi)^2} \quad (11)$$

The Average distance D_{avg} is

$$D_{avg} \frac{D}{2} \quad (12)$$

The members of the cluster will have the distance greater than the D_{avg}, and the election of the cluster head is now based on the maximum residual energy in the node and the ideal distance of the node.

C. *Securing the Data*

This protocol not only considers the residual energy and the ideal distance to form the network, but also secures the data transmitted between inter and intra clusters. To achieve the secured data, the message digest algorithm used. In message digest, there is a variation such as MD2, MD4, and MD5. The structure is same but MD2 designed for 8-bits machines, but MD4 and MD5 designed for 32-bit or higher machines. The MD4 is faster than MD5 but it is not an absolute secured one. The MD5 algorithm used to generate the 128-bit message digest from the input claimed to be as unique as fingerprints. The MD5 generates a hash value, which is unique for each of the input, that is, no two inputs can have the same hash value and the hash value generated is a one-way hash, the one-way hash means irreversible.

Table 2 Performance of the encryption algorithms

Hashing algorithm	*Key size/hash size (bits)*	*Extrapolated speed (Kbytes/sec.)*	*PRB optimized (Kbytes/sec.)*
TEA	128	700	–
DES	56	350	7746
Triple-DES	112	120	2842
IDEA	128	700	4469
RSA	512	7	–
SHA	160	750	25,162
MD5	128	1740	62,425

In this protocol, we are using the MD5 hash value to secure the data what we are transmitting in intra or inter cluster. The sink node on receiving the data generates the hash value and compares with the hash value generated by the sending node. If the hash values generated are same, then the data is accepted by the sink, otherwise it rejects the data. If the node continues to send the tampered data to the sink, then that node will be removed from the network and the same is informed to the cluster head that the node is compromised and should be left of the network.

Table 2 shows the performance comparison of the various encryption algorithms. The two-speed measurements in the figures are taken as rough estimates of the performance of the algorithms on a 330 MHz Pentium II processor in 1999. The "Extrapolated Speed" was based on the C source code published in 1996 with no particular attempt to optimize the code, whereas the "PRB" figure is the result of a substantial research effort to produce proprietary, which optimized implementations of the algorithms in assembly language [8–11]. As we can see, MD5 performs much faster than all other algorithms in both cases.

From Table 2, we can observe that the MD5's performance is better than the other hashing algorithm. We have used the MD5 hashing algorithm to secure the data.

4 Performance Analysis

In this section, we evaluate the SCDIR protocol against the CBER protocol using the NS2 simulator. The simulation parameters are shown in Table 3. The analysis we have done on the packet delivery ratio (PDR) and on the residual energy.

The protocol takes care of the active attack such as *compromised nodes*, if the node is compromised then it tampers the data and forwards it to the other node, that is, inter or intra cluster node. When the node receives the data, the data is tampered the node simply drops the packet.

The protocol also takes care of the vampire attack as this tampered packet can move around the network and deplete the energy of the node, but as the data is

Table 3 Network parameters used for simulation in NS2

Parameter	CBER	SCDIR
Sensing region	1000 m x 1000 m	1000 m x 1000 m
Initial energy	100 J	100 J
Transmission energy	0.005 J	0.005 J
Receiving energy	0.003 J	0.003 J
Minimum energy	0.005 J	0.002 J
Cluster Area (approximately)	250–300 m	250–300 m
Number of nodes	100	100

secured and packets are dropped as soon as the data is tempered it can not move around the network and deplete the energy.

The PDR is the number of packets sends by the number of packets received.

Figure 2 shows that when simulation time is minimum, both CBER and SCDIR will perform almost the same. But when time increases, the PDR decreases in SCDIR but in CBER it is more, because in SCDIR it is required to calculate the hash value for each of the data received and compared and then forwarded if it did not match then the packet is dropped due to this the PDR is low in graph better at the security.

Figure 3 shows the energy consumption of the CBER and the SCDIR of the network. The SCDIR consumption is less when compared to the CBER. The SCDIR perform the hash value for the data and then transmit over the network if data is tampered then that will be rejected by that node and it will not be further send which, in turn, saves the energy in the network. The energy is better in SCDIR because it takes care of the isolated node in the network.

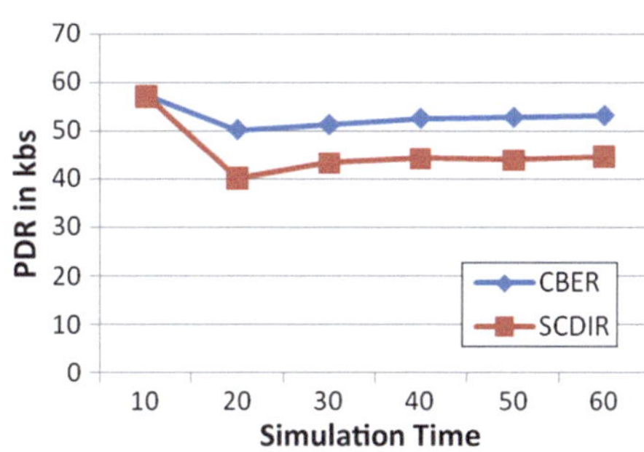

Fig. 2 The packet delivery ratio for the CBER and the SCDIR

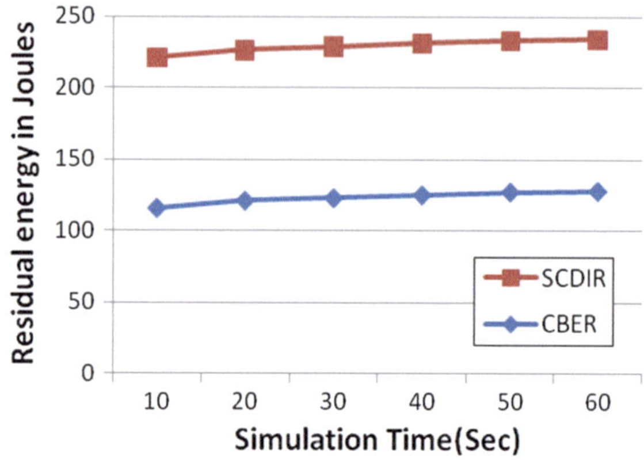

Fig. 3 The graph showing the residual energy for CBER and SCDIR

5 Conclusion

This paper proposes the SCDIR protocol, the protocol that calculates the residual energy of the nodes in the cluster and it finds the ideal distance between the member node and the cluster head periodically. The irregular shapes of clusters formed, which in turn eliminate the isolated nodes, and as the number of nodes participating in the cluster is more than the lifetime of the nodes that are prolonged in the network. The data are transmitted from the cluster members to the cluster head, without security, as the cluster head is aware of the member nodes and their locations in the cluster; it, in turn, removes the overhead of the nodes and save the energy. The hash value of the data, is generated which is sent to the sink along with the data to ensure the data integrity. This work is further enhanced to check that there is more than one node, which can become a cluster head in any rounds, or is any member node which is registered to more than one cluster to avoid the duplication of data or to avoid the false positive node.

References

1. Ranjani, S., Kavitha C.: "*Secured data integrity routing for sensor networks*", IEEE, International Conference on Advances in Electronics, Computer and Communication (2014)
2. Aboobeker. S.K.M. et al.: "*A Novel cluster-based energy efficient routing in wireless sensor networks*", Internation Conference on advanced Information Networking and Applications-2013, pp. 41–47
3. G. Kannan, T. Sree Renga Raja: "*Energy efficient distributed cluster head scheduling scheme for two tiered wireless sensor network*", Egypt. Inform. J. (2015)
4. Mohammad H., Robert N.: "*Adaptive routing in wireless sensor networks: QoS optimization for enhanced application performance*". Information Fusion xxx (2013) xxx–xxx
5. Santar Pal Singh, Sharma, S.C.: "*A Survey on cluster based routing protocols in wireless sensor Network*", ICACTA-2015, pp-687-695

6. Farzad K., Ehsan A., Mazdak A., Touraj K.: "*Efficient, intelligent energy routing protocol in wireless sensor networks*". Hindawi Publishing Corporation International Journal of Distributed Sensor Networks, Volume 2015, Article ID 618072, PN-13 pages
7. Jenq-Shiou Leu, Tung-Hung Chiang, Min-Chieh Yu., Kuan- Wu Su.: Energy efficient clustering scheme for prolonging the lifetime of wireless sensor network with isolated nodesIEEE Commun. Lett.192259262
8. Songhua Hu, Jianghong Han, Xing Wei, Zhen Chen: "*A multi-hop heterogeneous cluster-based optimization algorithm for wireless sensor networks*". Springer Science + Business Media New York 2014, Wireless Netw 21: 57–65 (2015)
9. Teng Gao, Jin-Hua Ding De-Quan Wang, Jin-Yan Song, Zhen-Yuan Si.:"*Clustering algorithm based on fuzzy weight for wireless sensor networks"*, International Conference on Mechatronics, Electronic, Industrial and Control Engineering (MEIC 2015)
10. Gaurang R., Madhuri B.: "*Improving energy estimation based clustering with energy threshold for wireless sensor networks*". International Journal of Computer Applications (0975 8887) Volume 113—No. 19, March 2015
11. George C., Jean D., Tim K.: *Distributed systems concept and design* 3rd edition, pp. 287–289 (2000)

Objects Detection and Tracking Strategies

Sureshbabu Matla and Ravi Subban

Abstract Object detection and tracking play a vital role in several applications like face detection, gait detection, vehicle detection and pose detection, and object recognition. The first step in the object detection algorithm is detecting the presence of any object in video images. The process of the object detection is accomplished using Hidden Markov Models, Support Vector Machines, Machine Learning Techniques, Pattern Recognition, Statistical Method, Scale Invariant Feature, structured visual dictionary (SVD), AdaBoost, Clustering Method, Bayesian Framework, Particle Filtering, Vector Quantization, and Feature Extraction, etc. This paper presents a comprehensive study and analysis on object detection and tracking techniques.

Keywords Support vector machines · Machine learning techniques · Statistical method · Scale invariant feature · AdaBoost

1 Introduction

In the past forty years, the detection and tracking an object in motion pictures of a surveillance system is attracting more and more interest among the researchers due to its wide range of applications. The object tracking plays an important role in the detection of terrorists in public places such as bus stands, airports, railway stations, universities, and office buildings. So, the use of CCTV cameras in public places is an unavoidable act, which, of course, requires spending some money. Even, some of the residential places like apartments and individual house require the installation of CCTV cameras to track the entry of unknown people like thieves and terrorists. These cameras typically span large spatial areas and do not have overlapping fields

S. Matla (✉) · R. Subban
Department of Computer Science, Pondicherry University, Pondicherry, India
e-mail: suresh.matla@gmail.com

R. Subban
e-mail: sravicite@gmail.com

S. Urooj and J. Virmani (eds.), *Sensors and Image Processing*, Advances in Intelligent Systems and Computing 651, https://doi.org/10.1007/978-981-10-6614-6_17

of view (FOV) is an improved cover. Object detection and tracking have other application of monitoring the loved ones like old parents and young children staying in homes.

Object tracking gives you more freedom in knowing that you are connected to your loved ones almost all day and from anywhere with tracking the information from the website or with the mobile phones. This can be done by global positioning system (GPS), which requires an object tracking device to be installed in the target place where the person you wish to locate and keep on watching. Automatic analysis of large videos significantly improves the quality of the monitoring [1].

Chau et al. [2] proposed a novel parameter tuning approach for tracking of an object. Usually, the tracking performance depends on video scene conditions such as illumination, density, and occlusion level of an object. The contextual features are computed for every frame, and the contextual feature is represented by a codebook called feature codebook, classified as clusters and the parameters are tracked. Peng et al. [3] presented a model based on motion boundary of video analysis. The bag-of-features framework is used for video-based human action recognition with the accuracy of 95.6%.

Cárdenes et al. [4] presented a novel algorithm with an idea of finding the shortest distance between the two points in 2D space. This was achieved by Locally Nearest Hidden Pixels. The concept used is analogous to the bucket sort algorithm. For validating this method, some of the important parameters were considered with respect to accuracy and computational load. As compared to the four different algorithms, this technique has highest accuracy rate and computational complexity.

The rest of the paper is organized as follows. In Sect. 2, the research work done by other researchers in object detection and tracking techniques is given. The analysis of the results is shown in Sect. 3. The conclusion is given in Sect. 4.

2 Objects Detection and Tracking Methods

2.1 Bayesian Framework

Boccignone et al. [5] demonstrated a novel framework using Bayesian framework for face and body detection from video streaming. Sometimes, cascaded classifier has no probabilistic interpretation. To solve this problem, the probabilistic interpretation of the output is embedded in a Bayesian tracker called boosted tracking (BT), which is very good with respect to generality and computational efficiency.

2.2 Image Processing-Based Methods

Ari et al. [6] used a new algorithm for image edge detection using ant colony optimization algorithm and Fisher ratio. The movements of ant in the image were determined using a probability matrix direction, and heuristic information of possible directions was determined. F-ratio-technique determined the optimum threshold of updated pheromone matrix. Experimental results show that the proposed method is efficient and accurate, and the performance of this algorithm is better as compared to other edge detection methods such as Canny, Sobel, Prewitt, Log, and Roberts. Bodor et al. [7] presented a novel approach for human motion and gait recognition systems. The existing systems have used a single camera for capturing images, which provided input that is view dependent. But the classification accuracy of this algorithm is poor. This process captures the input from multiple cameras and digitizes the input into the frames and computes the position of each frame and finally uses an image-based rendering to create orthogonal views as training and test data producing good recognition rate 93% on indoor scenes.

2.3 AdaBoost Learning Algorithm

Stojmenovic [8] proposed a novel-based AdaBoost learning algorithm to build a fast and reliable object detection strategy in images based on small training sets. This algorithm provides a strong classifier by incremental addition of weak (WC) on the basis of demonstrated eliminate functions whose best threshold near the trivial position is at the minimum or maximum threshold of all values producing the detection rate of 98.7% and 0.4% false positive rate. This approach not only recognized cars, but it can be used in many object tracking applications.

2.4 Statistical Method

Jeffrey et al. [9] presented a statistical tracing algorithm for tracking pedestrian and vehicle traffics. The evaluation was performed based on the intensities for predicting the traffic intensities. The statistical tracker produced results comparable to those of the MHT. The ability of statistical method is to easily use multiple cameras, and it is the main advantage and suited for tasks such as estimating vehicular traffic in large areas where data association for conventional tracking is not practical.

Proença [10] provided an algorithm with an aim of comparing the performance of the four most popular algorithms, SIFT, SURF, BRIEF, and DAISY, based on the detection and matching strategies. Especially DAISY and SURF methods have greater impact on performance than the matching distances. SIFT algorithm was in one extreme with the power range of high matching distances. BRIEF algorithm is not the most effective method for this application. Besada et al. [11] used a novel algorithm for the identification of video aircraft of tail number recognition, which is related to the airport real time images and calculates joint probability of each tail number. The major aims of this system are providing the accuracy and robustness in tracking all objects of interest.

2.5 Support Vector Machines

Teoh [12] explored a vehicle detection system based on vehicle edges and symmetrical characteristics. This method used two-class classification algorithm, which is an edge-oriented histogram (EOH) feature extractor and a support vector machine (SVM) for classifying vehicles. Once a vehicle is verified, vehicles are tracked based on Kalman filter. The results showed that this method is able to detect 94.6% vehicles appearing in the video frames. Glasner et al. [13] demonstrated a view-specific support vector machines method for level detection and viewpoint estimation for rigid 3D objects from single 2D images. This method relies on a nonparametric representation of a joint distribution of shape and appearance of the object class.

2.6 Scale Invariant Feature Transform

Bastanlar et al. [14] described a pipeline for structure from motion (SfM) with mixed camera types. Scale invariant feature transform technique is used for matching the automatic feature points between omnidirectional and perspective images and a weighting strategy is used for estimating camera poses. Based on these experimental results, it is inferred that the proposed method has better performance in matching an image.

2.7 Structured Visual Dictionary (SVD)

Yang et al. [15] proposed an algorithm by incorporating geometric into the visual vocabulary referred as structured visual dictionary (SVD), and it is used for enhancing the strength between the foreground object and the background. For the experimental results, image sequence frames were used, and the geometric information helps to improve the tracking performance, in particular, when the target object undergoes occlusion and large pose change. This algorithm produces low-average tracking errors.

Croon et al. [16] presented a method that compares the active vision models with two different approaches, namely probabilistic approach, and behavioral approach. The probabilistic models are classified into four different types: action selection based on mutual information (MI), learning for entropy reduction (L-ER), action selection for most probable class (MB) and learning for classification performance (L-CP) and a comparative study is made. Among the models, L-CP only yields the good performance in different actions run with different inputs.

2.8 Pattern Recognition

Polak et al. [17] demonstrated a method for evaluating the performance of image segmentation algorithms by novel error measure, which is identified while segmenting the multiple objects in an image based on object by object. Error measure defined number of mis-segmented pixels in an image. Among these error measures and OCE yield less error rate while segmenting the image. Fuh et al. [18] presented an algorithm for pattern recognition which is projected at different angles. This algorithm works based on the steps: image generation, normalization process, weighted projection, and matching process. In the first step, the image is taken as input then using the equivalent method the given image is rotated, and the vertical projection is found, if the pattern size is larger than image, it is put into larger frame. Variation of size leads to the difficulty in matching the image. To avoid this, the normalization process is performed.

Darrell et al. [19] proposed a novel approach for real time tracking of a person in crowded environment through the integration of multiple visual modalities, the depth measurement, color segmentation, and intensity pattern classification. Tracking was performed at three different time scales: short range, medium range, long range. Ayers et al. [20] introduced a system for automatically recognizing human actions from video sequences taken from a room. This system starts with prior information of layout of the scene, and it is a combination of low-level components skin detection, tracking and scene change and high-level components recognition. Several sequences were tested with this system, and it is capable of recognizing all interested human actions.

2.9 The Clustering Method

Xu et al. [23] presented a multi-target visual tracking framework for object recognition and tracking with background subtraction. This method solves fragmentation problem. The background-matching method is applied to determine whether two blob sets showing coherent motion are really the same object or different objects. The quantitative and qualitative experimental results show that the clustering method can group blobs properly. The performance of the background adaptation process is also satisfactory. Song et al. [24] proposed an algorithm for solving the problem of defect detection in random textures using color and texture information of an image.

2.10 The Particle Filtering

Khan et al. [25] used a novel approach for tracking an object using the combination of particle filters and multi-mode anisotropic mean shift. Input image is supplied as input to the particle filtering step, and the shape of the object is estimated and the appearance is allocated by mean shift. The probability/likelihood of an object of bounding box is estimated. Online learning process is used for updating the reference object distribution. Tracking of an object is robust in term of long-term partial occlusions and intersections of objects.

2.11 Vector Quantization

Li et al. [26] presented a transformed vector quantization (TVQ) algorithm for achieving highest compression ratio of selected images, which combines the features of code book and vector quantization (VT). Image compression techniques are used to reduce the data rate based on channel capacity. Image compression techniques are classified into lossless and lossy compression techniques. Lossless compression is error free. In this method, the image is transformed to produce efficient code representation then quantization is performed to reduce the number of output symbols. This stage has the significant effect in the compressed image.

Idris et al. [27] presented a method based on vector quantization (VQ) for indexing of compressed images and video. Large database systems require efficient indexing to enable fast access to the images in a database. This technique combines compression and indexing. The average retrieval rate is 95%.

2.12 Feature Extraction Method

Rittscher et al. [28] proposed a new feature extraction method for the classification of human motion, which performs automatic segmentation of sequences containing mixed motions. Spline contour is used to track the outline of the person at different positions. An autoregressive process (ARP) is most suitable model for tracking the contour dynamically. The condensation filtering is used for switching automatically between different motion classes.

2.13 Machine Learning Techniques

Vel et al. [29] presented a novel approach based on classification of lines subtending an object of 2D image views for high-recognition rates with reduced computational time and one more issue to be considered is that 3D objects need to generate basic model and perform matching scheme which leads to more difficult and proposed model simplifies this task. This approach is compared with other machine learning techniques based on recognition-by-parts-and-relations (RPR) paradigm. For this purpose, object data base is used.

3 Discussion

In this section, we summarize the object detection and tracking methods, point out limitations, and identify promising directions for future research to address these limitations. Kalman filter techniques have proven to yield good results, and they can usually be extracted features with low computational cost. However, their applicability is limited to all the scenarios reliably. Moreover, they cannot deal with occlusions. Generative state-space models such as HMMs can model temporal variations but have difficulties in distinguishing between related actions; discriminative graphical approaches are more suitable. The current state of the art models and well-known models motivated by the broad range of applications that can benefit from robust to detection of moving objects in videos. It is expected that many of these challenges will be addressed in the near future (Tables 1 and 2).

Scale invariant feature transform Y. Bastanlar et al. [14] described a pipeline for structure from motion (SfM) with mixed camera types. Scale invariant feature transform technique used for matching the automatic feature points between omnidirectional and perspective images, and a weighting strategy is used estimating camera poses in hybrid pairs [22]. Experiments conducted on fish eye images for matching results of all hybrid pairs SIFT technique gives false positive (FP) 0.30 and true positive (TP) 56.2 and proposed method yields 0.02 FP rate and 74.0 TP rate.

Table 1 The Performance Analysis of object Detection and Tracking Techniques

Author name	Methodology used	Merits	Demerits	Success rate/accuracy
Bedagkar-Gala et al. [1]	Automated video surveillance	High performance	Scalability issues	
Chau et al. [2]	Codebook method	Trajectories successfully tracked		80%
Peng et al. [3]	Bag-of-features framework	Reduce the influence of the camera movement		95.6% to KTH, 87.6% to youtube51.8% to HMDB51
Cárdenes et al. [4]	Locally Nearest Hidden Pixels	Highest accuracy, reduce the computational complexity		
Boccignone et al. [5]	Bayesian framework. BOOSTED TRACKING	Computational efficiency		
Ari et al. [6]	Ant colony optimization and Fisher ratio	Efficient and accurate detection		
Bodar et al. [7]	Human motion and gait recognition systems	Captures the input from multiple cameras and digitizes the input into the frames	The classification accuracy of this algorithm is poor.	93% on indoor scenes.
Stojmenovic et al. [8]	AdaBoost learning algorithm	High-recognition rates		98.7% and 0.4% false positive rate
Boyd et al. [9]	MHT	Reduce error rate	Practical impossible	
Proença et al. [10]	Matching strategy	Matching of features are good	Depends on environment	
Besada et al. [11]	Aircraft identification algorithm	Correct identification of images		83%
Teoh et al. [12]	EOH and SVM	Ambiguity reduced		94.6%
Glasner et al. [13]	Support vector machines	Good precision and accuracy rates		99.16% and 85.28%
Bastanlar et al. [14]	Structure from motion	Performance is good		0.02 false positives and 74 true positives
Yang et al. [15]	Structured visual dictionary	Low tracking errors		

Table 2 The Performance Analysis of object Detection and Tracking Techniques (Contd.)

Author name	Methodology used	Merits	Demerits	Success rate/accuracy
Croon et al. [16]	Probabilistic approach	Good performance in different actions runs with different inputs.		
Polak et al. [17]	Image segmentation	Using novel error measure	Error measure defined number of mis-segmented pixels	
Darrell et al. [18]	Depth, color, and intensity Pattern classification	High-performance results		100% and 87%
Darrell et al. [19]	Depth, color, intensity Pattern classification	High-performance results		100% and 87%
Han et al. [21]	Hidden markov models and MAP	Higher precision rate in event classification		98.3%
Xu et al. [23]	Multi-target visual tracking framework	Solve fragmentation problem	Not efficient complicated situations	
Li et al. [26]	Transformed vector quantization	Good response time		
Idris et al. [27]	Vector quantization	Retrieval rates are good		95% and 94% respectively
Rittscher et al. [28]	Autoregressive process, Condensation filtering	Eliminate tracking problems		
Vel et al. [29]	Lines subtending an object of 2D image views	High-recognition rates with reduced computational time	3D objects need to generate basic model and perform matching scheme	

4 Conclusion

This paper provides a comprehensive review of object detection and tracking algorithms and uses the existing developments and various types of video surveillance systems that are used for object tracking, motion analysis, behavior understanding, and recognition. The inspiration for writing a survey paper on this topic is to evaluate and achieve insight into visual surveillance systems from a big picture first. This allows us to understand what are the trends and the various stages of a visual surveillance system.

References

1. Bedagkar-Gala, A., Shah, S.K.: A survey of approaches and trends in person re-identification. Image Vis. Comput. **32**(4), 270–286 (2014)
2. Chau, D.P., Thonnat, M., Brémond, F., Corvée, E.: Online parameter tuning for object tracking algorithms. Image Vis. Comput. **32**(4), 287–302 (2014)
3. Peng, X., Qiao, Y., Peng, Q.: Motion boundary based sampling and 3D co-occurrence descriptors for action recognition. Image Vis. Comput. **32**(9), 616–628 (2014)
4. Cárdenes, R., Alberola-López, C., Ruiz-Alzola, J.: Fast and accurate geodesic distance transform by ordered propagation. Image Vis. Comput. **28**(3), 307–316 (2010)
5. Boccignone, G., Campadelli, P., Ferrari, A., Lipori, G.: "Boosted Tracking in Video," 17 (2):129–132 (2010)
6. Ari, S., Ghosh, D.K., Mohanty, P.K.: Edge detection using ACO and F ratio. SIVP **8**(4), 625–634 (2013)
7. Bodor, R., Drenner, A., Fehr, D., Masoud, O., Papanikolopoulos, N.: "View-independent human motion classification using image-based reconstruction." Image Vis. Comput. 27 (8):1194–1206 Jul. (2009)
8. Stojmenovic, M.: Real time machine learning based car detection in images with fast training. Mach. Vis. Appl. **17**(3), 163–172 (2006)
9. Boyd, J.E., Meloche. J.: "Evaluation of statistical and multiple-hypothesis tracking for video traffic surveillance," pp. 344–351 (2003)
10. Proença, H.: "Performance evaluation of keypoint detection and matching techniques on grayscale data," SIVP. no. 3, Aug. (2013)
11. Besada, J.A., Molina, J.M., Garc, J., Berlanga, A., Portillo, J.: "Aircraft identification integrated into an airport surface surveillance video system," pp. 164–171 (2004)
12. Teoh, S.S., Bräunl, T.: Symmetry-based monocular vehicle detection system. Mach. Vis. Appl. **23**(5), 831–842 (2011)
13. Glasner, D., Galun, M., Alpert, S., Basri, R., Shakhnarovich, G.: Viewpoint-aware object detection and continuous pose estimation. Image Vis. Comput. **30**(12), 923–933 (2012)
14. Bastanlar, Y., Temizel, A., Yardimci, Y., Sturm, P.: "Multi-view structure-from-motion for hybrid camera scenarios," Image Vis. Comput. vol. 30(8), 557–572 Aug. (2012)
15. Yang, F., Lu, H., Yang, M.-H.: Learning structured visual dictionary for object tracking. Image Vis. Comput. **31**(12), 992–999 (2013)
16. de Croon, G.C.H.E., Sprinkhuizen-Kuyper, I.G., Postma, E.O.: Comparing active vision models. Image Vis. Comput. **27**(4), 374–384 (2009)
17. Polak, M., Zhang, H., Pi, M.: An evaluation metric for image segmentation of multiple objects. Image Vis. Comput. **27**(8), 1223–1227 (2009)
18. Fuh, C., Liu, H.: "Projection for pattern recognition," 16(2) 1997, pp. 677–687 (1998)
19. Darrell, T., Gordon, G., Harville, M., Woodfill, J.: "Integrated person tracking using stereo, color, and pattern detection," vol. 37, no. 2, pp. 175–185 (2000)
20. Ayers, D., Shah, M.: "Monitoring human behavior from video taken in an office environment," vol. 19 (2001)
21. Han, M., Xu, W., Tao, H., Gong, Y.: Multi-object trajectory tracking. Mach. Vis. Appl. **18**(3–4), 221–232 (2007)

22. Shivappa, S.T., Rao, B.D., Trivedi, M.M., Fellow, L.: "Audio-visual fusion and tracking with multilevel iterative decoding : Framework and experimental evaluation," vol. 4, no. 5, pp. 882–894 (2010)
23. Xu, W., Zhao, Q., Gu, D.: Fragmentation handling for visual tracking. SIVP **8**(8), 1639–1649 (2012)
24. Song, K.Y., Kittler, J., Petrou. M.: "Defect detection in random colour textures," vol. 14, pp. 667–683 (1996)
25. Khan, Z.H., Member, S., Gu, I.Y., Member, S., Backhouse, A.G.:"Robust Visual Object Tracking Using Multi-Mode Anisotropic Mean Shift and Particle Filters," vol. 21, no. 1, pp. 74–87 (2011)
26. Li, R.Y., Kim, J., Al-shamakhi, N.: "Image compression using transformed vector quantization," vol. 20, pp. 37–45 (2002)
27. Idris F., Panchanathan, S.: "Image and video indexing using vector quantization," pp. 43–50 (1997)
28. Rittscher, J. Blake, A., Roberts, S.J.: "Towards the automatic analysis of complex human body motions," vol. 20, pp. 905–916 (2002)
29. Vel, De O., Aeberhard, S.: "Object recognition using random image-lines," vol. 18, pp. 193–198 (2000)

Road Accident Informer System

S. Vanitha

Abstract When a road accident happens, the victims are taken to hospital after a delay. Because of this kind of delay, a number of people die on the way to the hospital or on the spot. This system will rectify this problem. When an accident occurs, this device will send a text message to the mobile number of the nearby ambulance service. It will also send a message to the police control room, nearby government hospital, and a nearby private hospital.

Keywords Accident · Sensors · Message

1 Introduction

As there is an increase in the sales of personal vehicles, especially the sales of cars, it leads to the large number of accidents every day. Approximately more than one million people die every year because of the road accidents all over the world. India is one of the topmost countries with the highest number of deaths because of the road accidents. The number of people got affected by injuries may cross more than 20 million every year. The road accidents not only affect the victims but also affect their family members. A large number of families were affected by the road accidents. It also results in economic losses to victims, their families, and the nations as a whole.

If the accidents happen in a remote highway and if no one is available nearby, then it may be delayed to inform, get an ambulance, and inform the police. Because of this delay in sending the victims to the hospital, a large number of people die on the spot or on the way to the hospital.

This new system, "Road Accident Informer System," will rectify this problem. A device can be designed to send information about the accidents to a nearby ambulance service, the highway patrol, control room, nearby government hospital, nearby private hospitals, and nearby police station.

S. Vanitha (✉)
Dhanalakshmi College of Engineering, Chennai, India

S. Urooj and J. Virmani (eds.), *Sensors and Image Processing*, Advances in Intelligent Systems and Computing 651, https://doi.org/10.1007/978-981-10-6614-6_18

After receiving the information, the ambulance will rush to the spot and the unnecessary delay will be avoided. This device can be used for the vehicles such as car, van, bus, truck. Even after getting the information, sometimes it may be delayed for the ambulance to arrive at the spot. Since the accident information simultaneously reaches the police station and highway patrol, they will check whether the ambulance reached the spot or not. If there is any delay, they will make necessary arrangements.

2 Proposed System

2.1 Working Principle

If an accident occurs, this device will automatically send information based on the vibrations. The vibration detection sensors will detect the vibrations continuously and send signals to the microprocessor-based system.

If the vibration exceeds beyond a particular limit, it will be detected by this system, the current location of the vehicle will be fetched from the GPS receiver, and the message will be sent to the ambulance services. Few vibration detection sensors should be mounted on the outside surface of the body of the vehicle.

Generally, two vibration detectors can be mounted in the front side, three along each side of the vehicle, and two at the rear end. Totally, ten sensors can be mounted on the body of the vehicle to detect the vibrations caused by the accidents.

For example, if a car hits another vehicle, only one side will get damaged. The vibration detector mounted on that side will get damaged. But the remaining vibration detection sensors will detect the vibrations and send signals to the microprocessor-based device which is mounted inside the vehicle. All the vibration detecting sensors will be connected to the "Road Accident Informer Device."

The vibration detectors will be controlled by the microprocessor controller. This device will be incorporated within a GPS system to fetch the current location of the vehicle and to send the text message to the ambulance service, etc.

If the device is mounted in front or at the rear end, there are a lot of chances to get damaged, i.e., when a vehicle hits another vehicle or hits another object. To avoid this problem, the device can be mounted inside the center of the vehicle. If the road condition is not good, that vibrations will be observed by the shock observers in all the vehicles.

When an accident happens, the vibrations will travel along the frame of the vehicle (or body of the vehicle). So the vibration detection sensors will detect this vibration and send a signal to the microprocessor controller (Fig. 1).

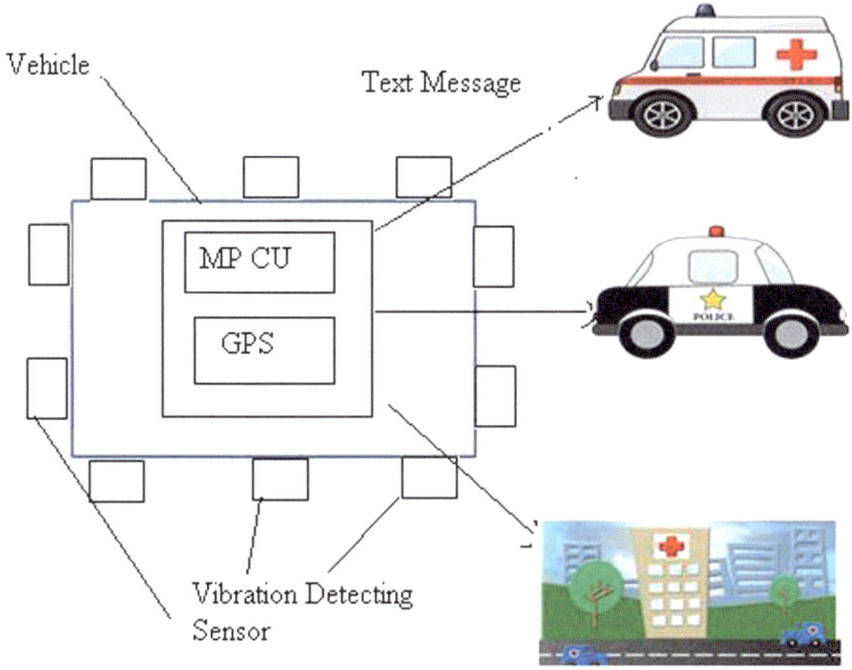

Fig. 1 Road accident informer system

2.2 Device Design

2.2.1 Sensors

Vibration sensors act as an initial source of information for this device. It is important to select the proper sensors to ensure reliable signal information. The moving parts in acoustic and vibration microsensors are especially susceptible to mechanical noise as a result of molecular agitation. In the sensors designed for small-signal applications, this mechanical–thermal noise is often one of the limitations.

Since a few decades ago, fiber optic sensors technology has experienced a revolution by the hand of fiber optic telecommunication product outgrows opto-electronic devices [1, 2, 3, 4, and 14].

Intensity-based sensor techniques have been studied and implemented in the last 25 years. A variety of configurations can be used, such as fiber microbending, fiber-to-fiber coupling, moving masks/gratings, and modified cladding [5–13].

Sensors are selected based on the frequency range. Since the vibration level will be very high during the collision of the vehicles and it is the primary source of information, it is very important to select appropriate sensors.

The Model 3028 vibration sensor is used in this system to detect the vibrations. It is a piezoresistive silicon accelerometer packaged on a ceramic substrate with a metal bracket which can be used to fix the sensor to the mounting location. The sensor consists of a micromachined silicon mass, and it is suspended by multiple beams from a silicon frame. The piezoresistors located in the beams generally change their resistance as the movement of the suspended mass changes the strain in the beams. Silicon caps are provided on the top and bottom of the device. This design is suitable for high shock resistance, durability, and built-in damping over a wide range of bandwidth. This 3028 vibration sensor seems to be appropriate to detect the vibrations in this new system. However, based on the performance, it may be replaced by a better sensor in the future in this device. A circuit can be designed which can detect and measure the vibration and shock created by the coalition of the vehicles. The sensors will be interfaced with the Arduino board general purpose input/output.

2.3 *Microcontroller*

The Mega 2560 Arduino board is used in this system. The memory capacity is high compare to other Arduino boards. More number of input/output pins are available in this model. This is the biggest and best Arduino. This model of the Arduino board is suitable for this system. A threshold value will be set. When the vibration signals cross the threshold value, an interrupt will occur. On the receipt of the interrupt, the controller momentarily stops and responds to the interrupt. The controller will start the execution of the particular interrupt service routine or interrupt handler. The microcontroller will fetch the location of the incident from the GPS receiver with the help of the interrupt handler. After fetching the location, using the GSM technology it will send the alert message to the nearby ambulance, police station, and government hospital. The mobile numbers will be fetched from the database. Since the Arduino board is used in this system, it reduces and simplifies the hardware and software requirements. The Arduino board has the power and circuit setup in order to program and communicate with the microcontroller over USB. The I/O pins of the microcontroller are already fed out to sockets/headers for easy access. A large number of libraries are available in the Arduino to make the programming easier.

2.4 GPS Receiver

The GPS receivers generally use the satellite-based Global Positioning System. The US Department of Defense operates the GPS, which depends on a worldwide network of satellites that constantly broadcast signals from space to the earth. The GPS receiver determines its position by merging the transmission data from four or more satellites.

GPS receiver is used in this system to fetch the current location of the vehicle. The accuracy of the GPS is generally around 30 feet (9 m). A WAAS-enabled (Wide Area Augmentation System) GPS unit can achieve position accuracy of 10 feet (3 m) under ideal conditions, particularly if it has a high-sensitivity chipset.

When an accident occurs, this system will automatically fetch the current location of the vehicle and then it will fetch the ambulance number, police control room number, and government and private hospital numbers in that nearby location and send the message immediately.

3 Experiments

The prototype model was created for this device and tested for cars. The results are given in Table 1. This device is more suitable for cars. It can be utilized in the buses by increasing the number of sensors (Table 1).

Table 1 Testing result for prototype model

Current location	Ambulance no.	Police patrol No.	Government hospital
Srirangam	108, 101	100	0431-2771465
	9688884980		

Few more tests will be conducted to prove the efficiency of this system. This system should be tested by using various vibration detection sensors in order to find an appropriate sensor to use in this system. The performance of this system is good in the highways. This system is yet to be tested in the remote interior roads.

4 Conclusion and Future Enhancement

This device can be used effectively during the road accident emergencies to send a message to the ambulance services and control rooms without any delay.

At first, this system can be implemented for highways and it can be extended to the state highways and all the other roads in the future. This device can be used for the two wheelers also by doing minor modifications. This system is yet to be tested for the remote interior roads. Since this device is fetching the location from the GPS receiver, the performance is based on the signals. This system should be tested in various locations to improve the performance.

References

1. Lequime, M.: Fiber sensors for industrial applications. In: Proceedings of the 12th International Conference on Optical Fibre Sensors, vol. 16 of OSA Technical Digest Series, pp. 66–71. OSA, Washington, DC, USA (1997)
2. Medlock, R.: Fibre optics in process-control. Control Instrum. **21**(4), 105–108 (1989)
3. Demjanenko, V., Valtin, R.A., Valtin, R.A.: A noninvasive diagnostic instrument for power circuit breakers. IEEE Trans. Power Delivery **7**(2), 656–663 (1992)
4. Dakin, J., Culshaw, B.: Fiber Sensors: Principals and Components, vol. 1. Artech, Boston (1988)
5. Giallorenzi, T.G., Bucaro, J.A., Dandridge, A., Siegel Jr., G.H., Cole, J.H., Rashleigh, S.C., Priest, R.G.: Optical fiber sensor technology. IEEE J. Quantum Electron. **18**(4), 626–665 (1982)
6. Hongbin, R., Jinlei, J., Zhishe, C., Dengyu, Z.: Application of spread spectrum technology to measurement. Air Force Institute of Missile Shaanxi, China (1997)
7. Perrone, G., Vallan, A.: A low-cost optical sensor for noncontact vibration measurements. IEEE Trans. Instrum. Measur. **58**(5), 1650–1656 (2009) (View at Publisher · View at Google Scholar · View at Scopus)
8. Culshaw, B.: Optical Fibre Sensing and Signal Processing. P. Pergrinus, London (1984)
9. Aydin, N., Arslan, T., Cumming, D.R.S.: A direct-sequence spread-spectrum communication system for integrated sensor microsystems. IEEE Trans. Inf. Technol. Biomed. **9**(1), 4–12 (2005)
10. Pitt, G.D., Extance, P., Neat, R.C., Batchelder, D.N., Jones, R.E., Barnett, J.A., Pratt, R.H.: Optical-fibre sensors. IETE Tech. Rev. **3**(8), 379–417 (1986)

11. Gunasilan, U.: Operative factors contributing to the selection of fiber-optic techniques for remote measurement of strain/stress. In: Proceedings of the IEEE 9th International Conference on Computer and Information Technology, Middlesex University Dubai (2009)
12. Hecht, J.: Understanding Fiber Optics. Pearson Prentice Hall, NJ (2006)
13. Gabrielson, T.B.: Mechanical-thermal noise in micromachined acoustic and vibration sensors. IEEE Trans. Electron. Devices, **40**(5), 903, 909 (1993)
14. http://www.hindawi.com/journals/js/2010/936487

Hybrid Intelligent Algorithm for Energy-Efficient Routing in WSN

Raminder Singh Uppal

Abstract Sensing data by sensor nodes in wireless sensor network (WSN) is random both in space and time. Routing of sensed data to base station in energy constrained WSN become more challenging as batteries of sensor nodes got consumed with every round of routing. Data packets are routed in multihop wireless communication in Time Division Multiple Access mode to base station. In this paper, soft computing techniques are used to propose an intelligent algorithm which enhances network lifetime by providing energy efficient routing. This is a hybrid approach in which genetic algorithm with partially mapped crossover is applied to find optimal routes while fuzzy logic is used to determine link cost. In order to make routing optimal, the link cost between adjacent nodes is calculated that consider residual energy of node, distance from base station, and density of nodes in a cluster. Fuzzy logic mechanism is used to calculate this link cost of all adjacent nodes, and these costs are represented in a link cost matrix which is updated after every round. This algorithm is based on hierarchal routing concept, and K-Mean numerical approach is used for clustering of sensor nodes. The approach is successfully implemented in MATLAB, and the simulation results of the various scenario show that the number of rounds before which the first node dies is more than LEEACH, thereby it enhances network lifetime as compared to LEEACH.

Keywords K-Mean · Soft computing · WSN · Routing

1 Introduction

Wireless sensors are considered to be minute, economical, and low power devices which are employed over an area in large quantities. Nodes integrate sensing units, transceiver, and actuators with limited on-board processing and radio capabilities

R.S. Uppal (✉)
Department of Electronics and Communcation, Baba Banda Singh Bahadur College of Engineering, Chandigarh Road, Fatehgarh Sahib 140407, Punjab, India
e-mail: rsuppal@gmail.com

S. Urooj and J. Virmani (eds.), *Sensors and Image Processing*, Advances in Intelligent Systems and Computing 651, https://doi.org/10.1007/978-981-10-6614-6_19

[1]. Each and every node is regarded as an inexpensive computer system which makes connection in addition to computation and also does the sensing [2]. Wireless sensor network is usually employed in realizing regarding the body parameter, creating arrangement supervisory, conditions supervisory, breach diagnosis, use of assistance program which can diagnose temperature, mobility, sound, and can sense earthquake [3].

Deployment of sensor nodes in target place is done often in random as well as deterministic method based on the application. After successful deployment, sensor must be capable of arranging themselves into wireless communication network. As a result of severe energy limitations involving deployed sensor nodes, the idea needs a selection involving various methods for management characteristics such as deployment, coverage, energy-efficient routing, and network security [4].

The standard routing methods includes a number of disadvantages whenever used on WSNs such as limited battery and minimal bandwidth of the wireless links provide routing challenges in WSN.

A. **Wireless Sensor Network Routing Protocol**

Routing in sensor networks differs from modern-day connection and also from wireless ad hoc networks in several means. It is not feasible to make a universal approaching system pertaining to the usage of large amount of sensor nodes. For this reason, IP-based approach of addressing could not be used for WSN routing [5, 7]. Base station sends request to all the deployed sensors, and this data-centric approach fits well to WSN Routing. It is challenging to design energy-efficient routing protocol for energy-constraint sensors where energy is consumed by sensor nodes in sensing, data processing, and routing the data to base station [6].

Routing techniques usually are grouped while using deployment structure straight into three categories of flat, location-based, and hierarchal routing protocol [5]. They have common objective of enhancing network lifetime and not compromising sensing data.

Within hierarchal routing, different functions and roles are usually assigned to any or all nodes. All nodes are usually actively playing distinct functions around the network. Network is divided into different cluster. Every cluster contains quantity of nodes and also nominates one node as a head of that cluster. Communication around the network is usually routed via nodes simply to the cluster head, and therefore, cluster head transmits it to base station.

Energy consumption of cluster head will be additional as it will be linked to transferring data to all cluster nodes to basic station. It is therefore suggested to occasionally revolve the cluster head responsibility to different nodes of that cluster to ensure even energy consumption. The data redundancy is reduced in hierarchal routing due to data aggregation because the data of the node in cluster is usually combined at cluster head. Selection of cluster head plays significant role in extending lifetime of the network.

Clustering is key parameter in WSN routing. It is done on the basis of location of the sensors nodes. Location to all the deployed nodes is usually known as they are embedded with GPS. K-Means [15] and fuzzy [13] are most widely used WSN clustering algorithms. K-Mean method is a numerical and non-deterministic technique used to partition the deployed nodes into K disjoint clusters and also determine cluster head of each cluster. In fuzzy clustering techniques, a node can belong to number of clusters simultaneously but with different degree of membership function. Both the techniques help in reducing the energy consumption of sensor network thereby increasing the network lifetime.

Problem formulation

Constraints on network resources especially limited energy of sensors nodes and their deployment in harsh environment like battlefield pose challenges in designing routing algorithm for WSN. During network operations, node energy gets depleted, and when it touches the threshold value, its sensing becomes faulty resulting in complex behavior of network.

This complex behavior and harsh environment consume node energy in unbalanced manner in which some nodes which are frequently involved die early as compared to other network nodes. Mathematical models would find it difficult to optimize consumption of energy of individual nodes and balance the consumption of energy of whole network simultaneously.

To address this problem of unbalanced energy consumption, optimal routing which involves traffic distribution over all the networks nodes in different rounds can be used. These optimal routes are based on link cost having residual and transmission energy as its parameters.

Computation intelligence algorithms such as GA, fuzzy logic, ACO, and neural network provide solution to such non-predictive complex behavior, and their optimized results would extend the life time of WSN.

B. **Soft computing for WSN Routing**

The productive utilization of energy within WSN can be bottleneck that will influences network overall performance as well as its life time. Optimization of data routing significantly affect the energy consumption of nodes and increases the time in which nodes get drained out of energy [8, 9]. Deploying intelligent and intellectual soft computing techniques such as fuzzy logic (FL), neural network (NN), evolutionary algorithm (EA) have been applied to find energy-efficient routes [10].

2 Route Cost Metric

In multihop WSN routing, each sensor node is responsible for both data sensing as well as routing. In order to find energy-efficient route location of the deployed nodes, node energy and location of the base station are the key parameters [11, 12]. With every round of routing, the residue energy of sensor node is consumed in sensing and routing the data packets to base stations. To find energy-efficient route, the fuzzy route cost-based approach has been applied (Fig. 1).

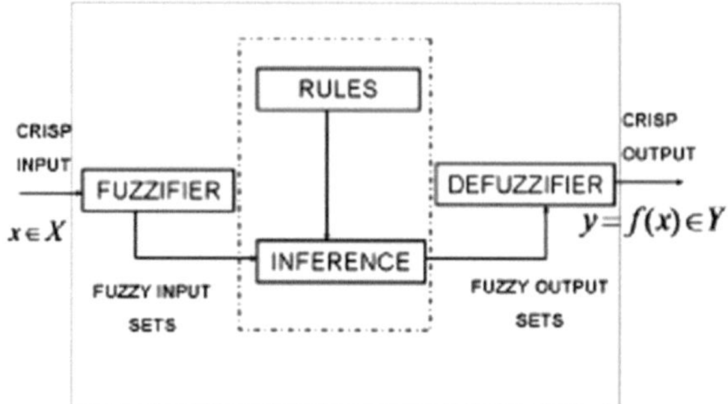

Fig. 1 Fuzzy system

Table 1 Fuzzy rule base

No	Distance(D)	Density(CD)	Energy(E)	Output
1	Med	High	V low	V long
2	Med	High	Low	long
3	Med	High	Med	med
4	Med	High	High	small
5	Med	High	V high	V small
6	Far	Low	V low	V long
7	Far	Low	Low	V long
8	Far	Low	Med	long
9	Far	Low	High	med
10	Far	Low	V high	med
11	Near	small	V low	V long
12	Near	small	Low	V long
13	Near	small	Med	med
14	Near	small	High	med

Fuzzy algorithm is based on the powerful capabilities of the fuzzy logic system to handle uncertainty and ambiguity. Fuzzy logic is well known as being model free. Their membership function is not based on statistical distribution. In this paper, we apply fuzzy logic system to optimize the routing process by some criterion. The main goal is designing the algorithm to use fuzzy logic systems to lengthen the lifetime of the sensor networks. Fuzzy system calculates the link cost between two sensor nodes which depends on residual energy, cluster weight, and distance of CH to base station. After every round of routing, energy of nodes involved in sensing and routing is consumed, and link cost is again calculated [13] (Table 1).

1. Residual Energy (E) [13]: Deployed sensor nodes are clustered by base station using K-Means method. Nodes after sensing sends data packets to respective cluster heads, cluster head receive data, and then forward it to base station in multihop communication mode. In a cluster mode, energy is consumed in

sensing data by nodes, cluster head consumes energy in receiving data from nodes, and energy spent in transmitting data from Ch to base station is given by Eq. 1. The residual energy of every node is calculated after every round of routing and given by Eq. 2.

$$\mathrm{Ec} = \sum_{J=1}^{M} E + \mathrm{M} *_{\mathrm{R}} + \mathrm{E}_{\mathrm{pz}} \tag{1}$$

$$\mathrm{E_{RES}}(\mathrm{r}-1) = \mathrm{E_{RES}}(\mathrm{r}-1) - \mathrm{E_c}(\mathrm{r}-1) \tag{2}$$

where r is the value of rounds in routing and $\mathrm{E_{RES}}$ is the residual energy.

2. Cluster Density (CD) [13]: Cluster density is the sum of distance from cluster head to nodes, CD for cluster having m member nodes would be

$$\mathrm{CD} = \sum_{i=1}^{m} \mathrm{d_{ip}}$$

where $\mathrm{d_{ip}}$ the sum of distance from i node to the cluster node p.

3. Distance (D): Distance of cluster head at location p from base station at location z would be defined as

$$\mathrm{D} = \mathrm{d_{pz}}$$

For a larger network, this distance should be minimized [14]; otherwise, the energy of most of the nodes will be wasted. However, for a smaller network that has a few closely located nodes, direct transfer to BS may be an acceptable option [17]. If residual energy is good and distance between CH and base station is near and distance between CH and nodes is high (density), then route cost is low.

Here residue energy, distance between CH and base station and density are inputs while route cost is output and near, good are linguistic variable of fuzzy system. This fuzzy system use Mamdani implication and trigonal membership function. There are 27 rules in rule base of this fuzzy system.

The proposed route cost measure consists of three parameters node energy, distance of base station, and cluster head and nodes. For a route between any two deployed nodes, route cost is calculated by fuzzy system. Cost of all possible routes among n nodes is stored in N*N route metric. After every round of routing, the fuzzy system calculates new values for the matric and this route metric cost changes.

3 Soft Computing-Based Routing Approaches

Energy of every node in network is consumed for sensing and routing the data to the base station. Those after every routing, round route cost metric of the network changes dynamically [18, 19]. The objective of this paper is to find out the energy-efficient routing policy which can enhance the network lifetime [10]. Soft computing approach suite well to ensure energy-efficient route selection among the routing nodes of the network [16].

a. **Partially mapped Crossover GA routing**

Optimization **Properties of Soft C**omputing approach of GA [13] is used to select energy-efficient routes among various nodes. The nodes of the sensors are analogous to genes of a chromosome and represented by numeric numbers, several chromosomes comprise a population of genetic algorithm, and the best chromosome generates the next population. Population transformation in future generation is depends on the survival fitness. Their performance is depending on the encoding scheme and the choice of genetic operators especially, the selection, crossover, and mutation operators.

GA pseudo code.

```
Choose initial population
 evaluate each individual's fitness
 determine population's average fitness
 repeat
      select best-ranking individuals to reproduce
      mate pairs at random
      apply crossover operator
      apply mutation operator
      evaluate each individual's fitness
      determine population's average fitness
 until terminating condition (e.g. until at least one individual has
      the desired fitness or enough generations have passed)
```

Population [13]: The initial set of the population is a randomly generated set of individuals. A new population is generated by two methods: steady-state GA and generational GA. The steady-state GA replaces one or two members of the population, whereas the generational GA replaces all of them at each generation of evolution.

Fitness [13]: In nature, an individual's fitness is its ability to pass on its genetic material. This ability includes traits that enable it to survive and further reproduce. In a GA, fitness is evaluated by the function defining the problem. The fate of an individual chromosome depends on the fitness value. The chances of survival are higher for better fitness values.

Selection [13]: The selection process determines which of the chromosomes from the current population will mate (crossover) to create new chromosomes. These new chromosomes join the existing population. This combined population will be the basis for the next selection. The individuals (chromosomes) with better fitness values have better chances of selection. We have used the rank selection method.

Crossover [13]: Analogous-to-sexual reproductive process in which genetic inheritance of parents is transferred to next generation children. Genes of chromosome of one parent got exchanged with the second parent to produce the gene of next generation. Permutation of n nodes provides us route, and route having minimum cost is the optimal solution. Conventional crossover would generate invalid routes with repeating and missing nodes. Partially mapped crossover is applied which provides only valid routes while moving form parent generation to next generation.

Pseudo Code of PMX

1. Randomly select a swath of alleles from parent 1 and copy them directly to the child. Note the indexes of the segment.

2.Looking in the same segment positions in parent 2, select each value that hasn't already been copied to the child.

A. For each of these values:

i. Note the index of this value in Parent 2. Locate the value, V, from parent 1 in this same position.

ii. Locate this same value in parent 2.

iii. If the index of this value in Parent 2 is part of the original swath, go to step i. using this value.

iv. If the position isn't part of the original swath, insert Step A's value into the child in this position.

3. Copy any remaining positions from parent 2 to the child.

Mutation [13]: If there is no significant improvement in survival of children after crossover, then new gene sequence is introduced in next generation. This process is called mutation. But selection process would only retain mutated chromosome if their fitness is higher than previous population involved in producing next generation. The inversion mutation is applied after PMX. This mutation process identifies two locations in a chromosome randomly and then interchanges the genes in substrings which lies between two locations.

4 Performance Evaluations

In the proposed method of energy-efficient routing, GPS-enabled sensors nodes are deployed in topographical area with the given dimensions. K-Mean clustering algorithm is applied which creates 10 disjoint clusters and also selects the head of each cluster [20].

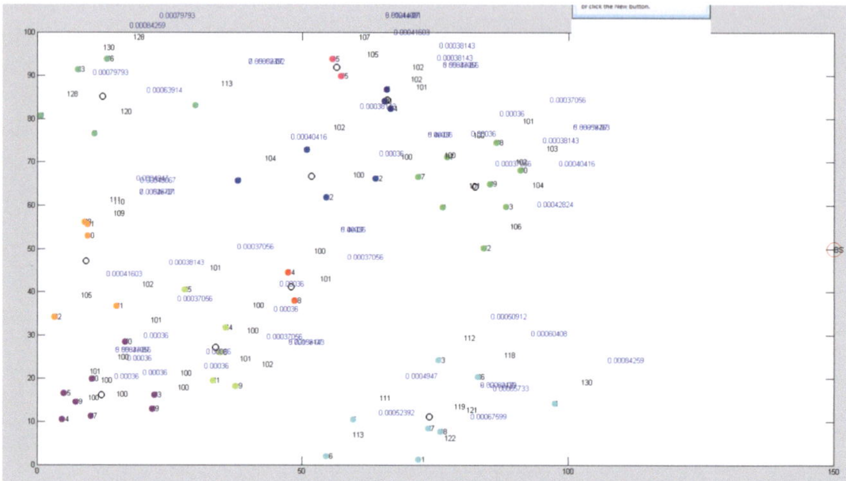

Population: Initial population of GA is generated with chromosome length of ten. Genes of this chromosome is randomly selected in such a way that only one member of each cluster finds place. This population represents feasible route for data packets, and every population would select new cluster head.

Fitness: Fitness of each population is calculated using link cost matrix generated by fuzzy system, and these populations are arranged in ascending order. Two fittest populations would be involved in crossover. Partially mapped crossover and inversion mutation are used as genetic operators for optimization of routing (Table 2).

Table 2 Three reading for each simulation scenarios of WSN

No of nodes	No of clusters	No of generations (GA)	Area (mtr)	Energy	BS area	No. of rounds covered before node dead
50	10	50	100	0.25	X = 150,Y = 100	643
	10	50	100	0.25	X = 150,Y = 100	630
	10	50	100	0.25	X = 150,Y = 100	638
	10	50	100	0.5	X = 150,Y = 100	1367
	10	50	100	0.5	X = 150,Y = 100	1044
	10	50	100	0.5	X = 150,Y = 100	1205
	10	50	100	1	X = 150,Y = 100	1507
	10	50	100	1	X = 150,Y = 100	1246
	10	50	100	1	X = 150,Y = 100	1768

Table 3 Parameters of full radio

Serial no	Parameters	Values
1	Number of nodes (m)	100
2	Network size (A)	50 x 50 m
3	Message size (k)	2000
4	Base location (z)	150,50
5	Cluster head Rate	10
6	Radio model	Full radio
7	Cluster head transmission	Single hop

Table 4 Packet and energy information

Primary energy	0.1j
Eelect	50 nJ/bit
Efs	10 pJ/bit/m^2
Dco	87 m
Eda	5 nj/bit/signal
Packet size	4000 bits
Calculation energy	50 nJ/bit
Send_near energy	Packet_Size*(Elect + Eda) + Packet_Size*Efs
Send_far energy	Packet_Size*(Elect + Eda) + Packet_Size*Emp*dis^4
Receive energy	(Packet_Size*(Elect + Eda)*0.05)
Combination & compacting	(Packet_Size*(Elect + Eda)*0.05)*0.05

Table 5 Result and Comparisons

	0.25 J	0.5 J	1 J
Our approach	522	1367	1768
LEEACH	394	932	2608

Simulation Parameter

The simulation was run 3 times, and at each time, the node's starting energy (0,25, 0,5 and 1 J) was different. K-Mean clustering of m nodes and subsequently dynamic cluster head selection by population generation of GA enhance the network lifetime. The best fitness value of the route is calculated after 100 generation.

Comparison of the proposed approach with LEECH on the basis of number of rounds taken before first node dies is tabulated below (Tables 3 and 4):

The Table below gives the number of rounds data packets would be routed before first node dies (FND) (Table 5; Fig. 2 and Graph. 1)

Graph. 1 Comparison based on number of dead nodes

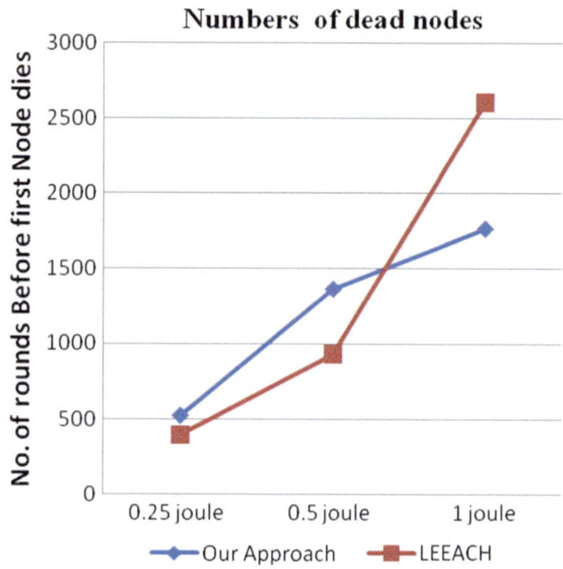

5 Conclusion

In this paper, dynamic changing conditions due to hostile deployment of sensors such as battlefield and energy depletion of sensor node are considered while designing energy efficient routing for wireless sensor network (WSN). A new hybrid approach of soft computing of fuzzy and genetic algorithm is proposed for routing. Fuzzy system calculates link cost among various nodes while genetic algorithm is used for finding efficient routes by considering link cost as fitness in this optimizing problem. Different scenarios are simulated in MATLAB, and result was compared with LEECH, and the results show improvement in terms of number of rounds before first node dies. More Soft computing approaches like ACO, BB_BC would be applied and also mobility of node would be considered in future work.

References

1. Akyildiz, IF., Su, W., Sankarasubramaniam, Y., Cayirci, E.: *Wireless sensor networks: A survey*. Computer Networks, 38(4), pp. 393–422, Mar (2002)
2. Sohraby, K., Minoli, D., Znati, T.: "Wireless Sensor Networks—Technology, Protocols, and Applications," Wiley, Inc, Publications (2007)
3. Haenggi, M., Ilyas M., Mahgoub, I.: „Opportunities and Challenges in wireless sensor networks", *Handbook of Sensor Networks: Compact Wireless and Wired Sensing Systems*, pp. 1.1 -1.14 2004: CRC Press

4. S. Misra et al. (eds.), *Guide to wireless sensor networks*, Computer Communications and Networks, doi:10.1007/978-1-84882-218-4 4, Springer-Verlag London Limited (2009)
5. Akkaya, Kemal, Younis, Mohamed: A Survey on routing protocols for wireless sensor networks. Ad Hoc Netw. **3**(3), 325–349 (2005)
6. Pantazis, N.A., Nikolidakis, S.A., Vergados, D.D.: Energyefficient routing protocols in wireless sensor networks: A survey. IEEE Communications Surveys and Tutorials **15**(2), 551–591 (2013)
7. Jamal Al-Karaki, Ahmed E. Kamal: "Routing Techniques in Wireless Sensor Networks: A Survey", *IEEE Communications Magazine*, 11(6), Dec 2004, pp. 6–28
8. Sharawi, M., Saroit, I.A., El-Mahdy, H., Emary. E.: „Routing Wireless Sensor Networks based on Soft Computing Paradigms: Survey." International Journal on Soft Computing, Artificial Intelligence and Applications (IJSCAI), 2(4), (2013)
9. Sajid Hussain, Abdul W. Matin, Obidul Islam: "Genetic Algorithm for Energy Efficient Clusters in Wireless Sensor Networks", *ITNG*, 2007, Information Technology: New Generations, Third International Conference on, Information Technology: New Generations, Third International Conference on 2007, pp. 147–154
10. Guo, W., Zhang, W.: A survey on intelligent routing protocols in wireless sensor networks. Journal of Network and Computer Applications **38**, 3–17 (2013)
11. Minhas, M.R., Gopalakrishnan, S., Leung, V.C.M.: "An online multipath routing algorithm for maximizing lifetime in wireless sensor networks," In Proc. IEEE Inform. Technol. New Generat. 6th Int. Conf., Apr. 2009, pp. 581–586
12. Azim M.A., Jamalipour, A.: "Performance evaluation of optimized forwarding strategy for flat sensor networks," In Proc. IEEE Global Telecommun. Conf., Nov. 2007, pp. 710–714
13. Chiang, S.Y., Wang, J.L.: Routing analysis using fuzzy logic systems in wireless sensor networks. Lecture Notes Comput. Sci. **5178**, 966–973 (2008)
14. Shengxiang Yang, Hui Cheng, and Fang Wang: _Genetic Algorithms With Immigrants and Memory Schemes for Dynamic Shortest Path Routing Problems in Mobile Ad Hoc Networks', IEEE Transactions on Systems, MAN, and Cybernetics—Part C: Applications and Reviews, 40(1), 2010, pp. 52–63
15. Hartigan, J.A., Wong, M.A.: J. Roy. Stat. Soc.: Ser. C (Appl. Stat.) **28**(1), 100–108 (1979)
16. Sharad, S., Shakti, K., Brahmjit, S.: " Hybrid intelligent routing in wireless mesh networks: Soft computing based approaches", 01, pp: 45–57
17. Shakti, K., Brahmjit, S., Sharad, S.: "Soft computing framework for routing in wireless mesh networks: An integrated cost function approach", 3, pp: 25–32
18. Tarique, H., Mariam, Y.: "A fuzzy approach to energy optimized routing for wireless sensor networks", 6, pp: 179–188
19. Heinzelman, W.R., Chandrakasan, A., Balakrishnan, H.: "Energy-efficient communication protocol for wireless microsensor networks," In *Proceedings of the Hawaii International Conference on System Sciences*, Jan 2000
20. Selim, B., Senol, Z.E.: "Genetic Algorithm Based Energy Efficient Clusters (GABEEC) in Wireless Sensor Networks", 10, pp: 247–254

A Method for Secret Image Transmission to Preserve Privacy

Rasika Thakare, B Sumit and Umesh Kulkarni

Abstract A proposed technique transforms mechanically a image, called as secret image into an another image so-called mosaic image, which carries same size data and also carries hidden data with it. The created mosaic image, which looks similar to randomly selected target image and also used as a cover up of the secret image. A image called as mosaic is produced by transforming the data and secret image onto the target image. To get back, the secret image, which is fixed into the created mosaic image, a technique is used called as data hiding, also uses a secret key.

Keywords Data hiding · Image encryption · Mosaic image · Secure image transmission

1 Introduction

Currently, electronic media use images for various purpose. The images, which hold secret information and have to protect them from misuse during transmissions [1–4]. In this paper, a system is proposed to transmit images securely, which transforms a image so-called secret image into a meaningful image called as mosaic image having the similar size as target image and carry the private text image. Secret key plays vital role in transformation. To store and recover the image called as secret image and secret data, receiver should have the secret key. The proposed method creates meaningful mosaic image and transforms an image having secret data into a mosaic image and also carry secret data with it.

R. Thakare (✉) · B. Sumit · U. Kulkarni
Kolhapur, India
e-mail: ranilpatil2011@gmail.com

B. Sumit
e-mail: b.sumit21@gmail.com

U. Kulkarni
e-mail: umesh.kulkarni@vit.edu.in

S. Urooj and J. Virmani (eds.), *Sensors and Image Processing*, Advances in Intelligent Systems and Computing 651, https://doi.org/10.1007/978-981-10-6614-6_20

2 System Design

The proposed method includes six main states as shown in diagram of Fig. 1:

1. Create Mosaic image.
2. Embed secret data.
3. Encrypt secret key.
4. Decrypt secret key.
5. Recover secret image using secret key.
6. Recover secret data using secret key.

2.1 Algorithm

- **Create cases**

1. To login into system.
2. To take inputs.
3. To merge two images.
4. To encrypt secret key.
5. To enter the secret data.
6. To hide secret data into image.
7. To decrypt secret key.
8. To recover secret image.
9. To recover secret data.

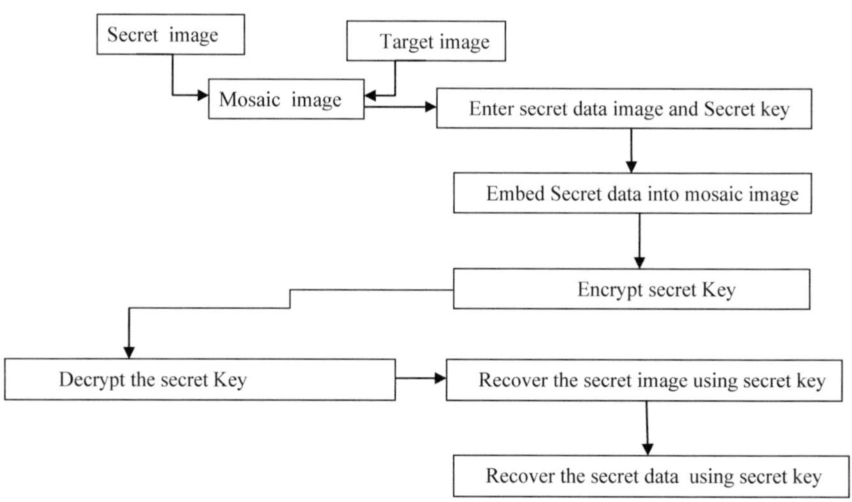

Fig. 1 system architecture

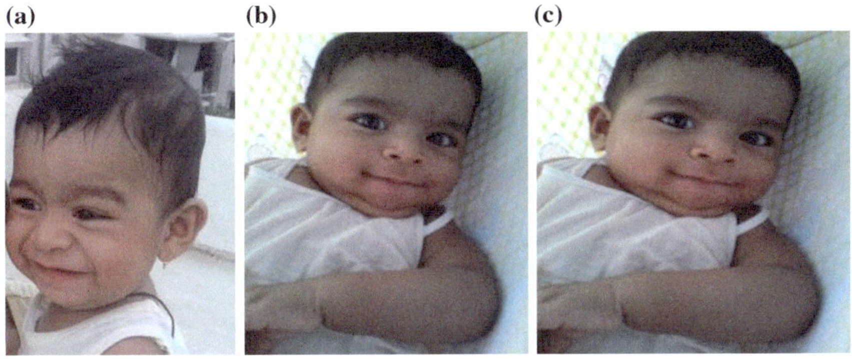

Fig. 2 (**a**) Secret image. (**b**) Target image. (**c**) Secret-fragment-visible mosaic image created from (a) and (b) by the proposed method

- **Case 1**

1. Enter userID.
2. Check user is sender or receiver.
3. If user == sender goto case 2 or case 7.

- **Case 2**

1. Enter secret key.
2. Enter two images as secret image and target image.

- **Case 3**

1. Read the two images as secret and target.
2. Calculate the size of each image.
3. Resize the target image to secret image and convert images to double.
4. Using the least significant bit technique merge two images and display it.

- **Case 4**

1. Read the secret key.
2. Take two prime numbers from database and using RSA algorithm encrypt the secret key using public key.

- **Case 5**

1. Enter the secret data from command prompt.
2. Convert the secret text into image and display it.

- **Case 6**

1. Read the two images as secret text image and merge image.
2. Calculate the size of each image.
3. Resize the merge image to secret text image and convert images to double.

4. Using the least significant bit technique merge two images and display it.

- **Case 7**

1. Using RSA algorithm decrypt secret key.
2. Display the secret key.

- **Case 8**

1. Enter secret key.
2. Read the merge image and covert the image to double.
3. Calculate the size of the image.
4. Using for loop convert the image to binary.
5. Using LSB technique recover the secret image and display image as secret image.

- **Case 9**

1. Enter key.
2. Read the merge image and covert the image to double.
3. Calculate the size of the image.
4. Using for loop convert the image to binary.
5. Using LSB technique recover the secret data image and display image as secret data image.

3 Result Analysis and Future Scope

A number of experiments have been conducted to test the proposed technique, using many secret and target images with different sizes, to explain that the created mosaic image which looks like the preselected target image.

3.1 Image Mosaicing Result Analysis

Here in Table 1, we have taken different secret images as well as target images. And then we have generated mosaic image. Here, we have calculated time required to create mosaic image and time required recover secret image from mosaic image.

In graph 1, comparison is shown. Time required to generate mosaic image as well as time required to recover secret image from mosaic image are shown. X axis represents images as secret and target image where Y axis represents time in seconds. In first entry, it is shown that mosaic image generation requires 21 s and recover secret image takes 18 s.

Table 1 shows image size(MB) v/s execution time (sec) to calculate image mosaicing and secret image recovery

NO	Name of secret image	Name of target image	Secret image size (MB)	Target image size (MB)	Time required for mosaicing (sec)	Time required for recovering secret image
1	Image1.jpg	Image2. jpg	225 × 300	206 × 276	21	18
2	Image3.jpg	Image4. jpg	600 × 800	2592 × 1944	68	64
3	Image5.jpg	Image6. jpg	2592 × 1456	2592 × 1456	92	88
4	Image7.jpg	Image8. jpg	1456 × 2592	1456 × 2592	72	68

Graph 1 shows Graphical representation of Image size on X-Axis v/s Time in seconds on Y-Axis for mosaicing of images as well as secret image recovery

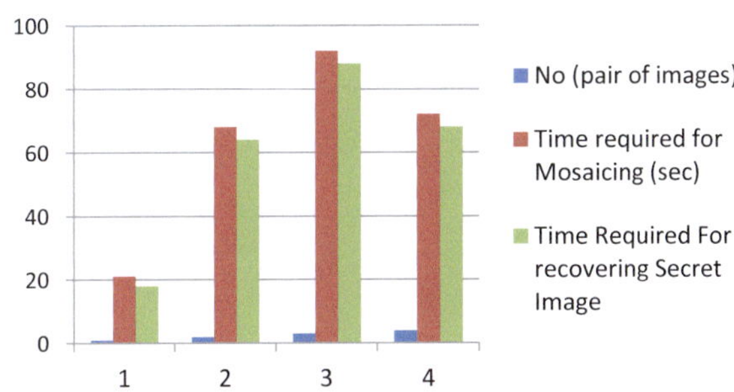

4 Conclusion

A new secure secret image transmission technique has been proposed, which can create meaningful image called as mosaic image as well as it can convert a secret image into a mosaic one with the same data size for use as a camouflage of the secret image and also carry hidden secret data with it. By the use of proper technique for image transformations, image called as mosaic image, which will look same as arbitrarily-selected image called as target image can be created with no need of a target image database. The original secret images can be recovered from the created mosaic images. Future studies may be directed to applying the proposed system to transmit secret data which will be in format as audio.

Acknowledgements The authors [1, 2 & 3] would like to thank Mr. Nilesh Patil for providing us with Fig. 2 (a–c).

References

1. Lee, Y.-L., Tsai, W.-H.: A new secure image transmission technique via secret-fragment-visible mosaic images by nearly reversible color transformations. IEEE Trans. Circ. Syst. Video Technol. **24**(4) (2014)
2. Lai, I.J., Tsai, W.H.: Secret-fragment-visible mosaic image—a new computer art and its application to information hiding. IEEE Trans. Inf. Forens. Secur. **6**(3), 936–945 (2011)
3. Chan, C.K., Cheng, L.M.: Hiding data in images by simple LSB substitution. Pattern Recogn. **37**, 469–474 (2004)
4. Coltuc, D., Chassery, J.-M.: Very fast watermarking by reversible contrast mapping. IEEE Signal Process. Lett. **14**(4), 255–258 (2007)

Text Retrieval from Scanned Forms Using Optical Character Recognition

Vaishali Aggarwal, Sourabh Jajoria and Apoorvi Sood

Abstract This paper investigates the use of image processing techniques and machine learning algorithm of logistic regression to extract text from scanned forms. Conversion of printed or handwritten documents into digital modifiable text is a tedious task and requires a lot of human effort. In order to automate this task, we apply the machine learning algorithm of logistic regression. The main components of this system are (i) text detection from the scanned document and (ii) character recognition of the individual characters in the detected text. In order to complete these tasks, we firstly use the image processing techniques to do line segmentation, character segmentation, and then ultimately character recognition. The character recognition is done by a one-vs-all classifier which is trained using the training data set and learns the parameters with the help of this data set. Once the classifier has learned the parameters, it could identify a total of 39 characters which include capital English alphabets, numerals, and a few symbols.

Keywords OCR · Logistic regression · Segmentation · Classifier

1 Introduction

In the real world, organizations or even individuals are faced with the problem where they have hard copies of certain documents and there is a need to convert the text in these documents to a digital form where operations like searching,

V. Aggarwal (✉) · S. Jajoria · A. Sood
Netaji Subhas Institute of Technology, Sector 3, Dwarka, New Delhi, India
e-mail: vaishalia809@gmail.com

S. Jajoria
e-mail: sourabhjajoria@gmail.com

A. Sood
e-mail: soodapoorvi@yahoo.com

S. Urooj and J. Virmani (eds.), *Sensors and Image Processing*, Advances in Intelligent Systems and Computing 651, https://doi.org/10.1007/978-981-10-6614-6_21

modification, insertion, [1] and others could be performed on the text. A system that could do this task is called an optical character recognition system (OCR) [2]. One important application of this system is to create a database from the information in filled forms which would otherwise be a very cumbersome task if done manually. The aim of the proposed model was to build an OCR that takes the scanned image of the document and accomplishes the aforementioned task. The OCR is required to be as accurate as possible. The most important step which determines the accuracy of the prediction is character recognition [3]. A one-vs-all classifier which uses a machine learning algorithm logistic regression [4–6] was used to build the OCR.

The idea behind one-vs-all classifier is that there were several classes which needed to be recognized. A training data set with objects of a number of known classes was used to train the classifier. The training data was composed of several entries where each entry had the values for a set of features along with the label for that entry. The classes in the case of an OCR were the set of characters that the system could recognize. For a given input character that had to be recognized, probabilities were computed using each of the classifier. Out of the resultant probabilities, the class for which the probability was the highest was given as the predicted class.

Section 2 describes OCR system design, Sect. 3 discusses the results of the experiments performed and a comparative study of the proposed model with other commercial OCR tools, and Sect. 4 concludes the paper providing some insight into the future applications of the proposed work.

2 OCR System Design

The main modules of the system [7–9] are described in this section.

a. Image acquiring, b. Pre-Processing, c. Line Segmentation, d. Character Segmentation, and e. Character recognition

2.1 Image Acquisition

In order to process a document [10, 11], firstly an image of the document was needed. The document in this case was a form which had only the text part and no images in it. The image had to be of good quality since the accuracy of the OCR depends highly on it. Noise had to be as low as possible. A scanner is the best device for this purpose. If the scanned image still contained some noise, then it had to be manually filtered out from it (Fig. 1).

Fig. 1 Scanned Form

REGISTRATION FORM

NAME VIMAL KUMARI

ADDRESS BUILDING NO 3 DWARKA SEC 14

CITY NEWYORK

ZIP CODE 3499712

AGE 56

CONTACT 34893366

EMAIL ID VKS25@LIVE.COM

2.2 *Pre-Processing*

Accuracy of an OCR system depends on text pre-processing and segmentation algorithms [12]. Pre-Processing can be further divided into the following steps:

- *The first step was to convert the colored image into a grayscale image, which consisted of shades of gray.*
- *Grayscale image was then converted into black and white image to filter out the unwanted noise.*
- *Small objects from binary image* [13] *were removed, i.e., objects that had fewer pixels than a cut-off value were removed from the black and white image. This step proved to be very efficient in the removal of noise* (Fig. 2, 3, and 4).

Fig. 2 Original Image

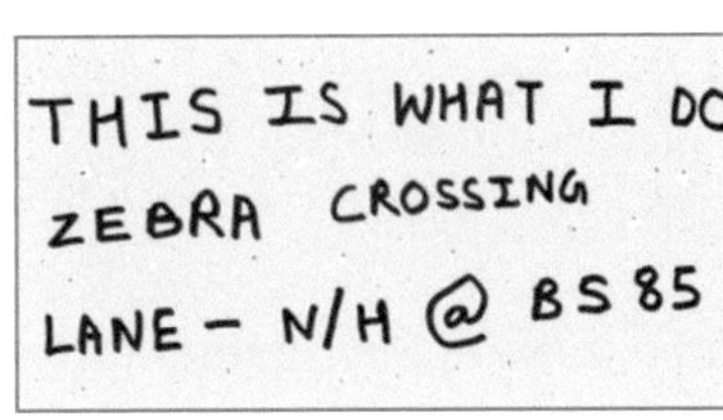

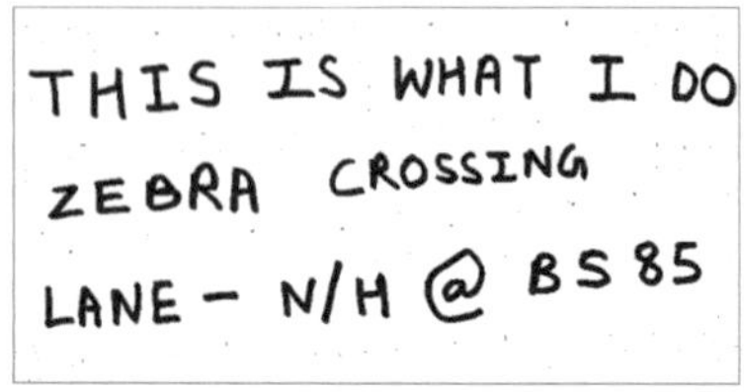

Fig. 3 Grayscale Image

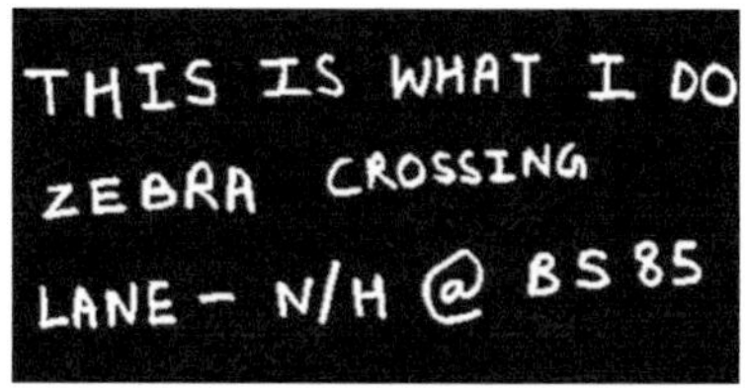

Fig. 4 Filtered Image

Fig. 5 Cropped Line

2.3 Line Segmentation

The image of the scanned document was to be processed line by line. Line Segmentation is the process of extraction of lines of text from the form and working on each line separately to facilitate character recognition in it. An important point to note down here is that the processing was based on inverted black and white images where the background is black with a pixel value of 0 and the written text is white with a pixel value of 1 [14, 15]. Steps applied for this purpose:

1. Starting from the top of the image, the row of pixels where the sum of pixel values was not zero was searched. This marked the beginning of the first line in the document.
2. The image was scanned till the sum of pixel values in that particular row was greater than zero.
3. The row where the sum of pixel values was zero marked the bottom of the current line. The aforementioned top and bottom pixel rows were then used to crop out a line from the image which was then used for character recognition.
4. The above three steps were repeated for the remaining image to get the remaining lines (Fig. 5).

2.4 Character Segmentation

Character Segmentation was used to obtain individual characters from the form field line obtained from line segmentation [16]. Each character was enclosed in a bounding box. To do this correctly, there needed to be some space between

Fig. 6 Characters in Bounding Box

Fig. 7 After top, bottom, left, and right cropping

characters. Now the bounded image of character was further cropped from all four sides of the box. This was done by scanning for the first line with non-zero pixel value sum from top, bottom, left, and right of the bounding box (Fig. 6 and 7).

2.5 *Character Recognition*

A one-vs-all logistic regression model was used to recognize the characters. Since there were 39 different characters that the system could recognize, 39 different classifiers were trained, one for each character.

A dataset of around 2000 handwritten and printed characters was used. This data was then used to train the classifier. Each element in the training data was a 24 * 42 pixel image of a printed or handwritten character (Fig. 8).

2.5.1 Logistic Regression

The logistic regression hypothesis is defined as:

$$h_\theta(x) = g\left(\theta^T x\right) \tag{1}$$

Here, g is the sigmoid function which is defined as:

$$g(z) = \frac{1}{1 + e^{-z}} \tag{2}$$

Fig. 8 Training Data samples

Θ is the parameter matrix whose value is different for all 39 classifiers. The dimensions of Θ for each classifier are 1 * (n + 1), where n is the number of features.

Logistic regression cost function:

$$J(\theta)=\frac{1}{m}\sum_{i=1}^{m}\left[-y^{(i)}\log\left(h_\theta\left(x^{(i)}\right)\right)-(1-y^{(i)})\log(1-h_\theta\left(x^{(i)}\right))\right]+\frac{\lambda}{2m}\sum_{j=1}^{n}\theta_j^2 \quad (3)$$

Where m is the total number of rows in the training data, x specifies each row of the training data and y specifies the label value for that particular row. The value of y varies from 1 to 39, one for each character. The second term in the above equation is the regularization term. Regularization was used to avoid the problems of over-fitting and under fitting. λ is the regularization constant.

Here, the cost function J(Θ) was to be optimized by changing the parameter Θ. For that, gradient of the cost function was calculated for each column in the Θ vector.

The gradient is a vector whose j^{th} element is:

$$\frac{\partial J(\theta)}{\partial\theta_j}=\frac{1}{m}\sum_{i=1}^{m}\left(h_\theta(x^{(i)})-(y^{(i)})x_j^{(i)}\right)+\frac{\lambda}{m}\theta_j \quad for\ 1\le j\le n \quad (4)$$

$$\frac{\partial J(\theta)}{\partial\theta_j}=\frac{1}{m}\sum_{i=1}^{m}\left(h_\theta(x^{(i)})-(y^{(i)}t)x_j^{(i)}\right)+\frac{\lambda}{m}\theta_j \quad for\ 1\le j\le n \quad (5)$$

This gradient function helped in determining the value of the parameter Θ for which the cost function was minimum. Thus, by finding the value of Θ, one-vs-all classifier was trained. In order to recognize a character, the probability that it belongs to each class using trained classifiers for each of the 39 characters was calculated. The label (1–39) for which the value of the probability was the highest was the character predicted by the one-vs-all classifier.

3 Results

The system was tested on several scanned images of forms which constituted the testing set. These forms contained printed fields and some handwritten data filled using ball-point or gel pen. The image was scanned using Microsoft Office Lens.

Figure 10 *shows the data stored in digital text format from recognition performed on* Fig. 9 *using our system.* The recognition rate for text on forms is between 85% and 90%, which is similar to other systems used in character recognition. There is still scope for improvement in recognizing similar characters like '0' and 'O', 'H' and 'B', and '1' and 'I'. Recognition rate depends a lot on these two factors:

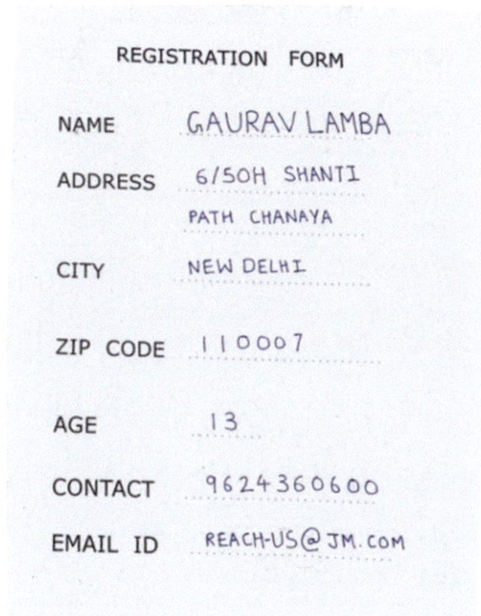

REGISTRATION FORM

NAME GAURAV LAMBA

ADDRESS 6/50H SHANTI

PATH CHANAYA

CITY NEW DELHI

ZIP CODE 110007

AGE 13

CONTACT 9624360600

EMAIL ID REACH-US@JM.COM

Fig. 9 Input Form

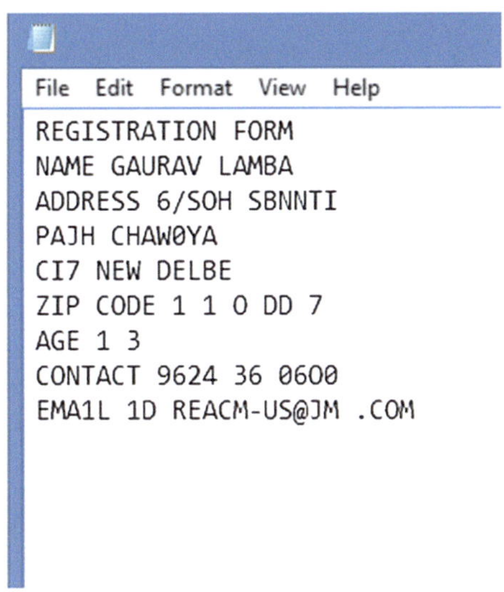

File Edit Format View Help

REGISTRATION FORM
NAME GAURAV LAMBA
ADDRESS 6/SOH SBNNTI
PAJH CHAW0YA
CI7 NEW DELBE
ZIP CODE 1 1 O DD 7
AGE 1 3
CONTACT 9624 36 0600
EMA1L 1D REACM-US@JM .COM

Fig. 10 Output

a. Quality of scanned forms and
b. Handwriting used on forms (Fig. 11).

The accuracy of character recognition was computed at different stages of development, and it was found that the accuracy of proposed OCR was proportional

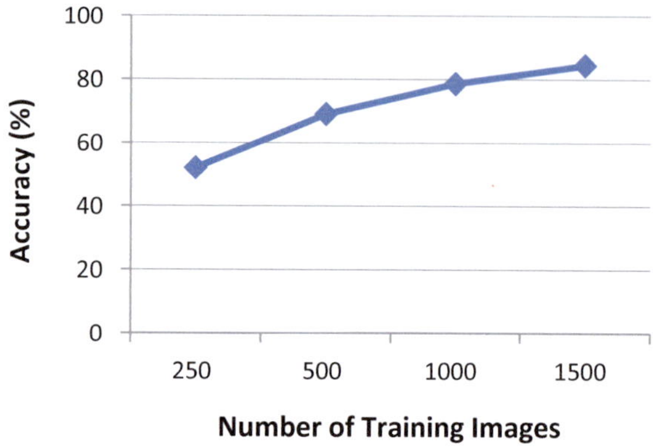

Fig. 11 Character recognition accuracy as a function of number of training images

Table 1 OCR Comparison

Input Image		Accuracy of OCR					
Image number	Number of characters	Suggested system	ABBYY FineReader 12	I2OCR	Tesseract	Cuneiform	Free OCR to word
1.	129	88.3	87.6	34.9	19.3	76.7	41.1
2.	128	93.7	85.2	42.2	57.0	58.6	64.4
3.	130	92.3	77.7	62.3	65.4	50.8	58.5
4.	125	79.8	73.4	52.6	59.9	51.2	55.3
5.	128	83.6	69.2	62.3	67.3	63.7	61.0
Average accuracy		87.4	78.6	50.9	53.8	60.2	56.1

to the number of training images in the training set at initial stages. The proposed system was also compared with some well-known free OCR tools like Tesseract, Cuneiform, ABBYY FineReader 12(Trial Version), I2OCR, and Free OCR to Word. They provide good accuracy and speed. Many other OCR tools are proprietary and paid. The accuracy of some tools was very less on the test set and hence they are not mentioned here. The experiment was carried out on a computer with Intel Core i3 1.7 GHZ CPU, 2 GB RAM, and Windows 8 OS. All tools were tested on the same scanned form images which were in the test set.

Table 1 compares all the relevant tools to extract text from scanned forms. The result of OCR processing shows that ABBYY FineReader 12 provides 78.6%, I2OCR provides 50.9%, Tesseract provides 53.8%, Cuneiform provides 60.2%, and Free OCR to Word provides 56.1% of average accuracy in the test set, whereas the suggested system provides an average accuracy of 87.4%.

4 Conclusion and Future Work

We tried to build a text recognition system that could be used on manually filled forms containing handwritten as well as printed English alphabets written in uppercase, numerals, and some special characters. There are numerous important domains of application of this system like library registration, banking forms, and many more. We have shown that text can be recognized with a reasonable accuracy using simple image processing techniques and logistic regression algorithm for multiclass classification of characters. It works well for a variety of documents having no prior knowledge of character size, font color, and document layout.

Future work in this system consists of modifying this approach to recognize lowercase letters and texts with different orientation. This system works for few special characters, so there is scope to add more characters. The machine learning algorithm used for classification is logistic regression; as an improvement, a more complex algorithm like neural networks can be used for better accuracy. Thus, one would hopefully obtain a more accurate recognition of filled forms to apply this system in practical use.

References

1. Mohammad, F., et al. (IJCSIT) International Journal of Computer Science and Information Technologies, **5**(2), 2088–2090 (2014)
2. Wolf, C., Jolion, M.J., Chassaing, F.: Text localization, enhancement and binarization in multimedia documents. In: International conference on pattern recognition, pp. 1037–1040, 2002
3. Kahan, S.T., Pavlidis, T., Baird, W.: "On recognition of printed characters of any font and size", IEEE transactions of pattern recognition and machine intelligence, pami-91987, pp. 274–285
4. Hosmer, D., Lemeshow, S.: Applied logistic regression, 2nd edn. Wiley, New York (2000)
5. Harrell, F.: Regression Modeling Strategies: With Applications To Linear Models, Logistic Regression, and Survival Analysis. Springer, New York (2001)
6. Logistic regression and artificial neural network classification models: A methodology review. J. Biomed. Inform. **35**, 352–359 (2002)
7. Jain, A.K., Bhattacharjee, S.: Text segmentation using Gabor filters for automatic document processing. Mach. Vis. Appl. **5**(5), 169–184 (1992)
8. An embedded application for degraded text recognition. EURASIP J. Adv. Signal Process. **2005**(13), 2127–2135 (2005)
9. Carson, C., Belongie, S., Greenspan, H., Malik, J.: Blobworld: Image segmentation using expectation-maximization and its application to image querying. IEEE Trans. Pattern Anal. Mach. Intell. **24**(8), 1026–1038 (2002)
10. Wolf, C., Jolion, J-M. Extraction and Recognition of Artificial Text in Multimedia Documents. http://rfv.insalyon.fr/wolf/papers/tr-rfv-2002-01.pdf
11. Doermann, D., Liang, J., Li, H.: Progress in camera-based document image analysis, in Proc. 7th IEEE International Conference on Document Analysis and Recognition (ICDAR'03), vol. 1, pp. 606–617, Aug 2003
12. Optical character recognition by open source OCR tool tesseract: a case study, Int. J. Comp. App. **55**(10), 0975–8887 Oct 2012

13. Matsuo, K., Ueda, K., Michio, U.: Extraction of character string from scene image by binarizing local target area. Transaction of The Institute of Electrical Engineers of Japan, 122-C(2), 232–241, Feb 2002
14. Gao, J., Yang, J.: An adaptive algorithm for text detection from natural scenes, in Proc. IEEE Computer Society Conference on Computer Vision and Pattern Recognition (CVPR '01), vol. 2, pp. 84–89, Kauai, Hawaii, USA, 2001
15. Sobottka, K., Bunke, H., Kronenberg, H.: Identification of text on colored book and journal covers, International Conference on Document Analysis and Recognition 57–63 1999
16. Chen, X., Yuille,A.: Detecting and reading text in natural scenes. In: Computer Vision and Pattern Recognition, vol. 2 (2004)

Combined Approach of Fuzzy Image Blocking Implementation for Medical Image Fusion

Madhuri Patel and Drashti Baldev

Abstract Image fusion is a procedure of creating a better quality image, which contains all important features of two or more input images. Evaluation of quality of output image is based on particular application. Currently, image fusion is performed using many techniques of transform domain and spatial domain methods. This research mainly includes implementation of medical image fusion using the fuzzy logic method. Moreover, this paper also presents that how we can improve the performance of medical image fusion by integrating fuzzy logic method with image blocking (IB). The result of the fusion process of the image is evaluated with quantitative parameters such as peak signal-to-noise ratio (PSNR), Entropy, and normalized cross correlation (NCC). Implementation result is presented in tabular and graphical form. From the derived results, it can be concluded that by integrating IB with fuzzy logic method performance gain can be achieved.

Keywords Image fusion · Fuzzy logic · Image blocking

1 Introduction

Several applications in the field of image processing need single image with high resolution [1]. However, due to the limitation of image capturing device, it is not possible to achieve high-resolution image with every small detail. Image fusion is helpful to overcome this limitation of image capturing device. The problem with capturing device is that lens can truly focus on objects that are at particular depth based on the focal length of the lens and remaining object, which are not in range of focal length will be out of focus and hence blurred [2]. The amount of blurring depends on variety of parameters like distance between object and lens, focal length

M. Patel (✉) · D. Baldev
AITS, Rajkot, Gujarat, India
e-mail: patel.madhuri1991@gmail.com

D. Baldev
e-mail: drashti.baldev@gmail.com

S. Urooj and J. Virmani (eds.), *Sensors and Image Processing*, Advances in Intelligent Systems and Computing 651, https://doi.org/10.1007/978-981-10-6614-6_22

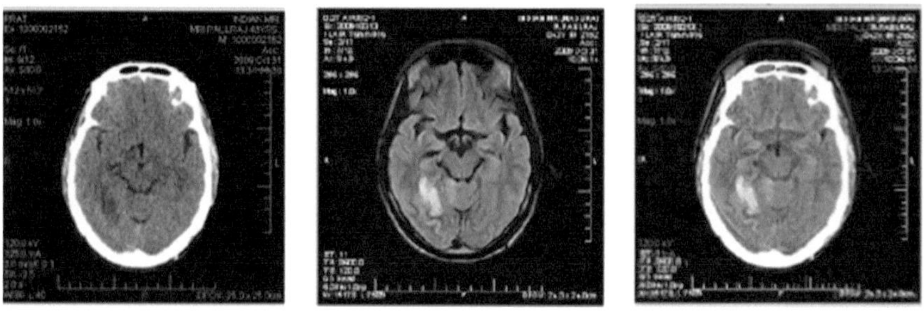

CT Image MRI Image Fused Image

Fig. 1 Example of Fused Image in Medical Field

of lens, and distance between sensor plane and lens. A well-focused image with every small detail can be achieved by means of image fusion. Image fusion can be applicable to many fields [2, 3], in this paper, we have covered the use of image fusion in the medical field. First two images in Fig. 1 are input images taken using computed tomography (CT) and magnetic resonance imaging (MRI), respectively [4, 5]. Hard structure of the body can be well focused in computed tomography images [6]. Soft tissues of the body can be well focused using magnetic resonance imaging [7]. By fusing these two images, we can achieve a single image with well-focused hard structure like bones and soft tissues both.

Better fused image will help us to gain countless medical benefits including better diagnosis, improved surgical planning, and accurate radiation therapy.

2 Implementation

2.1 Image Fusion Using Fuzzy Logic (FL)

Step 1: Take fuzzified input and set membership function. Gray scale input image consists of total 256 gray levels with pixel values ranging from 0 to 255. We have divided 256 gray levels into a fuzzy set made up of 5 membership functions such as follows: VL for very large; L for large; M for medium; S for small; and VS for very small. Resultant image will also contain 256 gray levels and uses the same fuzzy set.

Step 2: Generate fuzzy inference system. Basically fuzzy "IF-THEN" rules are generated according to fuzzy set classified in the first step.

Step 3: Pixels from both images are evaluated using fuzzy inference file generated in the second step.

Step 4: In this step, MOM(Min-sum Mean-of-max) defuzzification is performed.

2.2 Image Fusion with Image Blocking (IB)

Step 1: Get fused image using the fuzzy logic method mentioned above.

Step 2: Initial fused image F0, MRI image, and CT image are divided into square blocked with the equal size of m * n [8, 9]. Calculate the similarity measure SM values of the corresponding sub blocks of CT1, MRI1, and F01, respectively, using equations mentioned below [10–12].

$$SM_{F_0CT} = \frac{2 \times \sum_{x=1}^{m} \sum_{y=1}^{n} F_0(x,y) \times CT(x,y)}{\sum_{x=1}^{m} \sum_{y=1}^{n} \left[F_0(x,y)^2 + CT(x,y)^2\right]} \quad (1)$$

$$SM_{F_0MRI} = \frac{2 \times \sum_{x=1}^{m} \sum_{y=1}^{n} F_0(x,y) \times MRI(x,y)}{\sum_{x=1}^{m} \sum_{y=1}^{n} \left[F_0(x,y)^2 + MRI(x,y)^2\right]} \quad (2)$$

Step 3: Construct map using following equation:

$$\text{Map}\ (r,c) = \begin{cases} 1 & SM_{F_0CT}(r,c) \geq SM_{F_0MRI}(r,c) \\ 0 & \text{Otherwise} \end{cases} \quad (3)$$

Step 4: Generate the final output(fused) image F by the formula mentioned below:

If Map(m, n) = 1 and the total of each element in the Map entered at (m, n) of 3 × 3 neighborhood is equal to 9, then:

$$F(x,y) = CT\,(x,y) \quad (4)$$

Else if Map(m, n) = 0 and the total of each element in the map centered at (m, n) of 3 × 3 neighborhood is equal to 0, then:

$$F(x,y) = MRI\,(x,y) \quad (5)$$

Else

$$F(x,y) = F_0\,(x,y) \quad (6)$$

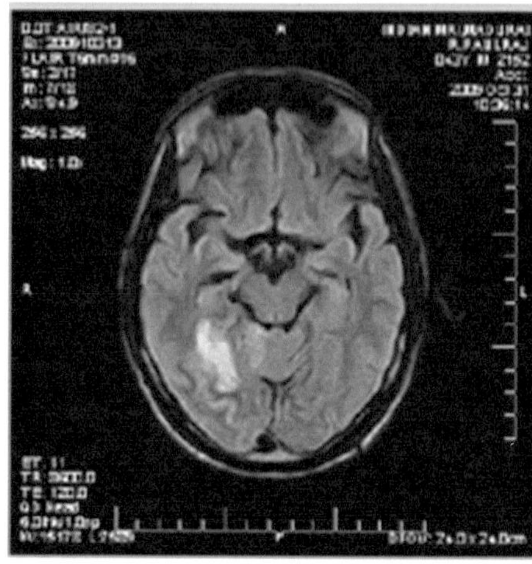

Fig. 2 MRI Image (Dataset2)

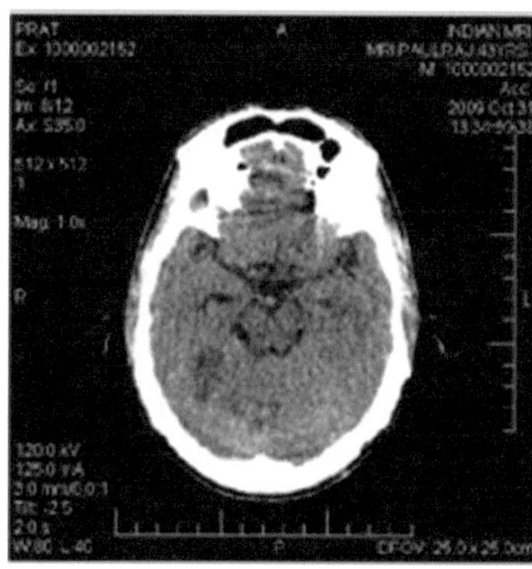

Fig. 3 CT Image (Dataset2)

2.3 Implementation Result

See Figs. 2, 3, 4, and 5.

3 Result Analysis

3.1 Evaluation Parameters

Quantitative parameters used in the evaluation of the proposed system are as following

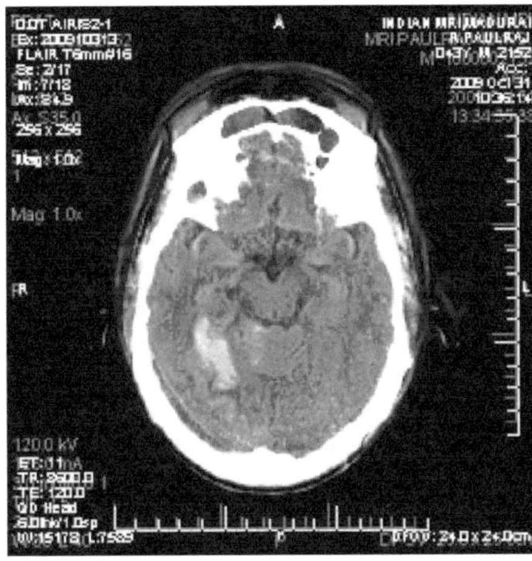

Fig. 4 Fused Image FL Method

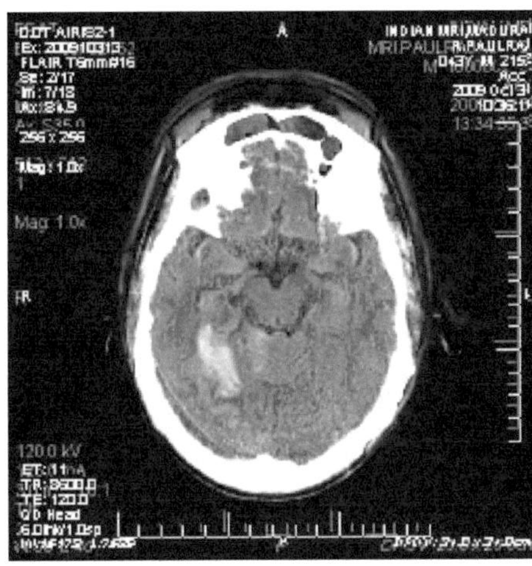

Fig. 5 Fused Image FL Method with IB

a) **Mean Square Error:** The root mean square error [3, 13] can be denoted by:

$$\sum_{x=1}^{m}\sum_{y=1}^{n}\frac{[R(\mathrm{x,y}) - F(\mathrm{x,y})]^2}{P * Q} \tag{7}$$

Where, R (x, y) is an original image (or one of the source images) and F (x, y) is the fused image. P and Q are the dimensions of the images. Better fusion performance can be indicated by small MSE.

b) **Peak signal to noise ratio:** Equation for peak signal to noise ratio [3] can be given as follow:

$$PSNR = 10 \ * \ \log_{10}\left(\frac{fm^2}{MSE}\right) \tag{8}$$

Where, denotes maximum gray scale value of pixels in final fused image, better fusion result can be denoted by higher value of PSNR.

c) **Entropy:** Entropy [2, 14] gives a statistical measure of randomness that can be used to characterize the texture of the input image. Entropy is defined as

$$E = -\sum_{x=0}^{L-1} P_x \log 2\, P_x \tag{9}$$

Where, p indicates count of histogram in output image.

d) **Normalized Cross Correlation:** Similarities between fused image and registered image can be obtained using normalized cross correlation(NCC). NCC can be calculated by the following equation [4]

$$NCC = \sum_{x=1}^{m}\sum_{y=1}^{n}\left(Ff_{xy} * I_{xy}\right) \Big/ \sum_{x=1}^{m}\sum_{y=1}^{n}\left(Ff_{xy}\right)^2 \tag{10}$$

Where, *I* is input image (or one of the source images) and *Ff* indicates final fused image. n, m denote the dimensions of the images.

3.2 *Quantitative Analysis*

See Figs. 6, 7, and 8; Tables 1 and 2

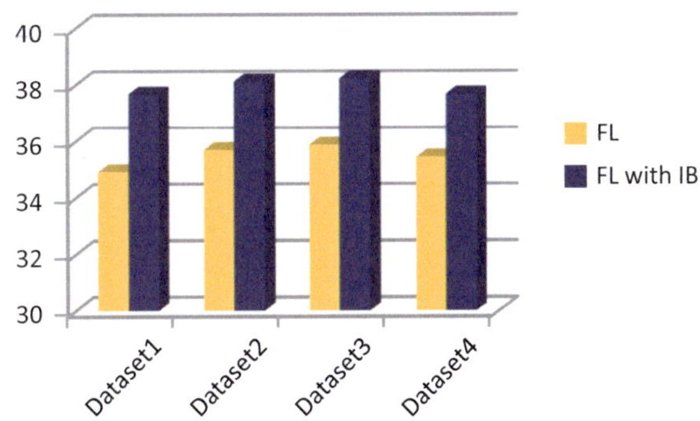

Fig. 6 Graph—PSNR using FL vs. FL with IB (Dataset1)

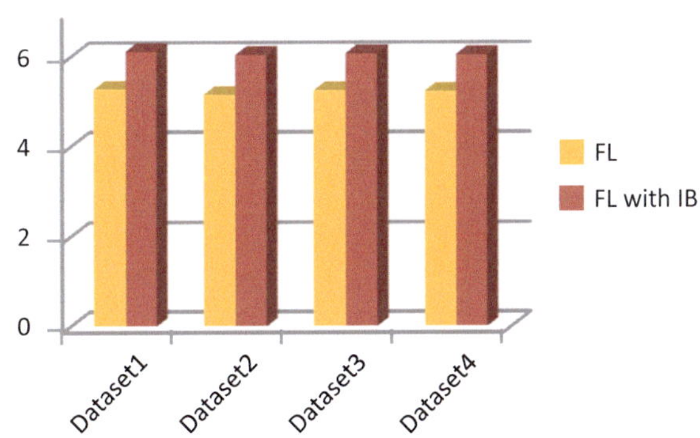

Fig. 7 Graph—Entropy using FL vs. FL with IB (Dataset1)

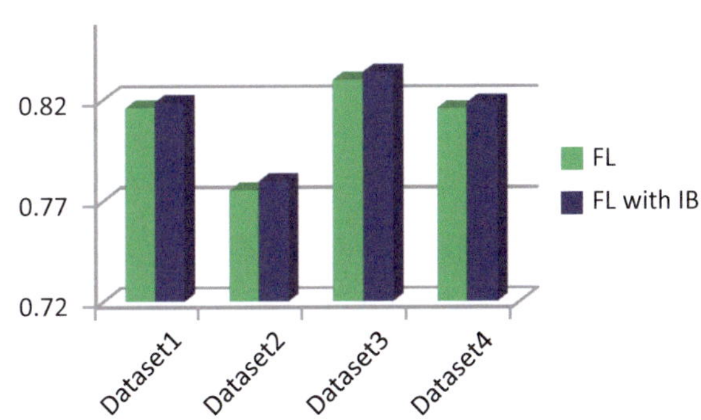

Fig. 8 Graph—NCC using FL vs. FL with IB (Dataset1)

Table 1 Quantitative analysis using FL and FL with IB methods

	Techniques					
	Fuzzy logic			Fuzzy logic with image blocking		
Parameters	PSNR	Entropy	NCC	PSNR	Entropy	NCC
Dataset1	34.93	5.2975	0.8159	37.69	6.1234	0.8187
Dataset2	35.69	5.1711	0.7752	38.13	6.0519	0.7798
Dataset3	35.88	5.2670	0.8300	38.24	6.0774	08339
Dataset4	35.44	5.2469	0.8158	37.69	6.0563	0.8189

Table 2 Quantitative analysis using FL method with various block size

	Fuzzy logic with image blocking								
	3 * 3			5 * 5			7 * 7		
Parameters	PSNR	Entropy	NCC	PSNR	Entropy	NCC	PSNR	Entropy	NCC
Dataset5	37.69	6.1234	0.8287	36.51	5.3614	0.8179	36.07	5.2898	0.8076
Dataset6	38.24	6.0774	0.8339	37.11	5.3561	0.8213	36.83	5.2593	0.8003
Dataset7	37.96	6.0563	0.8189	36.78	5.3478	0.8085	36.44	5.2539	0.7977

4 Conclusion

Several methods are available for image fusion. Spatial domain-based methods have disadvantages of reducing the contrast of the entire image. But as compared to transform domain methods, spatial domain methods are much useful in case of high contrast and bright images. Frequency domain approach can be used to overcome the limitations of spatial domain methods. Additionally, fuzzy logic is a good tool to deal with uncertainty.

So, we have implemented fuzzy logic with the integration of image blocking in which we can get improved result of fusion process as we have discussed above. Quantitative analysis is presented in the tabular and graphical form. Implementation result shows that by integrating fusion process with Image blocking better NCC, PSNR, and entropy values can be obtained. In this study when we are increasing block size, it improves the performance of fusion process deliberated above.

References

1. Zhang, Y.:Understanding Image Fusion, Photogrammetric Engineering & Remote Sensing, June 2008
2. Mohamed, M., El-Den, R.: Implementation of Image Fusion Techniques for Multi-Focus Images Using FPGA, 28th National Radio Science Conference, Apr 2011
3. Prakash, C., Rajkumar, S., Chandra Mouli, P.V.S.S.R.: Medical image fusion based on redundancy DWT and Mamdani type min-sum mean-of-max techniques with quantitative

analysis. In: 2012 International Conference on Recent Advances in Computing and Software Systems (RACSS), vol. 54, no. 59, pp. 25–27, Apr 2012
4. Desale, R.P., Verma, S.V.: Study and analysis of PCA, DCT & DWT based image fusion techniques. In: International Conference on Signal Processing Image Processing & Pattern Recognition (ICSIPR), vol. 66, no. 69, pp. 7–8, Feb 2013
5. Parmar, K., Kher, R.K., Thakkar, F.N.: Analysis of CT and MRI image fusion using wavelet transform. In: 2012 International Conference on Communication Systems and Network Technologies (CSNT), vol. 124, no. 127, pp. 11–13, May 2012
6. Hongbo, W., Yanqiu, X.: Pixel-based image fusion using wavelet transform for SPOT and ETM + image. In: 2010 IEEE International Conference on Progress in Informatics and Computing (PIC), vol. 936, no. 940, pp. 10–12, Dec 2010
7. Pradnya, P.M., Sachin, D.R.: Wavelet based image fusion techniques. In: 2013 International Conference on Intelligent Systems and Signal Processing (ISSP), vol. 77, no. 81, pp. 1–2, Mar 2013
8. Seng, C., Bouzerdoum, A., Tivive, F.: Fuzzy Logic-Based Image Fusion for Multi-View Through-the-Wall Radar, Digital Image Computing: Techniques and Applications
9. Rao, S., Seetha, M., Prasad, K.: Comparison of Fuzzy and Neuro Fuzzy Image Fusion Techniques and its Applications. International Journal of Computer Applications, Apr 2012
10. Ali, F., Dokany, I., Saad, A., Abd El-Samie, F.: Fusion of MR and CT Images Using The Curvelet Transform, 25th National Radio Science Conference, Mar 2008
11. Hamdi, M.: A Comparative Study In Wavelets, Curvelets And Contourlets As Denoising Biomedical Images, Image Processing & Communication, 2011
12. Do, M.N., Vetterli, M.: The contourlet transform: an efficient directional multiresolution image representation, IEEE Trans. Image Proces. **14**(12), 2091, 2106 (2005)
13. Patel, M., Brahmbhatt, K., Shah, Z.: Comparative Study on Image Fusion Methods. Int. J. Graphics Image Process. **4**(1), (2014)
14. Patel, M., Brahmbhatt K.: Medical Image Fusion. Int. J. Adv. Eng. Res. Dev. (IJAERD). **1**(5), (2014)

A Systematic Review on Image Enhancement Techniques

Nishant Bhardwaj, Gagandeep Kaur and Pradeep Kumar Singh

Abstract Image enhancement is a very important process used for enhancing the quality of an image, and for this purpose, numerous techniques have been used and applied. This paper describes how to enhance an image using different techniques. Various techniques have been studied for image enhancement used in different research papers. Understanding and reviewing the techniques used for image enhancement are the main goal of this paper. Our review work leaves a lot of future scopes which can be extended further by researchers.

Keywords Image enhancement · Image quality · Digital image processing

1 Introduction

Image enhancement is a process by which we can improve the quality of the digital image which makes it easier for identifying features. This can be done by removing noise, sharpening, or brightening an image. The techniques identified in the paper to enhance an image are contrast stretching and image sharpening, nonlinear image enhancement technique, genetic algorithm, generalized fuzzy enhancement, wavelet transform technique, multi-scale and single-scale retinex improvement technique, etc. This paper is divided into four sections. Section 1 is an introduction about the image enhancement followed by review procedure in Sect. 2. Section 3 covers the detailed analysis of the selected papers. Finally, in conclusion, answers to the research questions as well as the future scope are discussed in Sect. 4.

N. Bhardwaj (✉) · G. Kaur · P.K. Singh
Department of CSE, ASET, Amity University Uttar Pradesh, Noida, India
e-mail: nishant.bhardwaj92@live.com

G. Kaur
e-mail: ggndp262@gmail.com

P.K. Singh
e-mail: pksingh16@amity.edu

S. Urooj and J. Virmani (eds.), *Sensors and Image Processing*, Advances in Intelligent Systems and Computing 651, https://doi.org/10.1007/978-981-10-6614-6_23

Table 1 Research questions identified

Research questions
RQ1: To identify the various techniques used in the image enhancement?
RQ2: To compare the traditional image enhancement techniques with the new image enhancement techniques?
RQ3: To identify the factors affecting the image quality?

2 Review Procedure

To study the image enhancement techniques, we have read around 150 research papers. We have adopted a similar approach taken by Singh et al. [1, 2, 3] and then identified the most suitable papers using the following steps: (i) we have downloaded the papers using the keywords: image enhancement techniques, (ii) we read the abstract, title, and conclusion, (iii) we have chosen the research papers which are related to our study based on the content of papers, (iv) we have chosen 30 most relevant papers out of 150 papers, (v) the findings have been reported after studying each paper thoroughly, (vi) finally, we made tables which categorize the major contribution of authors, and techniques used in various papers. Following are the research goals to study the image enhancement techniques (Table 1).

3 Related Work

In this section, various papers related to image enhancement have been taken into account for review and analysis.

In [4], Negi et al. discussed contrast stretching and image sharpening techniques. It is an approach that concurrently adjusts contrast and enhances boundaries of an image. On the gray-scale image contrast stretching is applied and then it proceeds to Laplacian mask, and finally, Laplacian image is appended to the original gray-scale image to obtain the desired sharpened image.

In [5], Wu et al. proposed image enhancement using wavelet-based contourlet transform with cycle translation. In this, wavelet transform is used for decreasing the redundancy occurring in the original method of contourlet transform. WBCT and cycle translation are merged. At last, to magnify the images, adaptive enhancement function is selected. The proposed method can efficiently magnify the images and decrease the flecked at the background region, the image edges.

In [6], Wang et al. discussed NIE (Nonlinear Image Enhancement). Simulation and identification processes are used along with the proposed NIE method. This process uses clipping and scaling parameters which are an appropriate combination of various images. This process enhances the quality of blurred image and a better quality is achieved, and PSNR (signal-to-noise ratio) performance is obtained than other nonlinear enhancement techniques.

In [7], Sree et al. implemented the FPGA-based retinal fundus image enhancement algorithms. The design was accomplished using Simulink, and the implementation was accomplished on the Spartan 3E Development Board. It was found out that hardware system performance through different experiments was more as compared to software version for complicated video inputs.

In [8], Yaping et al. discussed the stepwise refinement method. The main objective of this method is how to recognize and enhance the particular portion of a digital image, and maintaining other information of the object by keeping it unchanged. So stepwise refinement method is used to figure out the problem. It comprises of image pre-processing, recognition and enhancement.

In [9], Imtiaz et al. discussed a color image enhancement method. FICE algorithm is used to enhance the RGB endoscopic images. The color reproduction is carried out by transferring chrominance map from the source image to the enhanced gray-scale spectral image in the procedure of corresponding luminance and texture information. It throws light on some of the tissue characterization on the surface part of the base endoscopic image.

In [10], Premkumar et al. discussed the color image contrast enhancement technique. Firstly, RGB image is transformed to HSV (hue, saturation, and value) color space. Hue color channel is selected for DST decomposition. The lower directional sub band is used for reformation. By transforming HSV to RGB color space, the enhanced image is acquired. From proposed DST-based contrast image enhancement approach, the satisfactory result is attained.

In [11], Bhattacharya et al. proposed a fast algorithm for raising the contrast of an image locally by using singular value decomposition (SVD) approach. The contrast of a partially degraded image is magnified by using this approach. Under global and partial degradation, the method is carried out properly and a good perceptual quality after processing is achieved.

In [12], Teng et al. described the Laplacian and Gaussian methods and discovered an upgraded Laplacian pyramid image enhancement method. On the obtained threshold of the Laplacian pyramid improvements, there is a good solution for the color difference between the smaller image detail enhancement and for the large difference in the color image enhancement, the effect is very obvious.

In [13], Wang et al. proposed an improvement to the wavelet transform method, naming it the lifting wavelet transform method for image enhancement purposes. It contrasts the new method with the existing wavelet transform method. The lifting wavelet transform method accredits the image to have wide gray-scale scope and also the detail is clear.

In [14], Munteanu et al. proposed real-coded genetic algorithm method for automatic image enhancement named EVOLEHA. In order to attain better explorative behavior, the method applies a real-coded GA with significant modifications like PCA-mutation. Due to a more exploitative crossover and selection scheme, the search is well-balanced and durable. EVOLEHA attains a combined goal (e.g., efficiency, robustness, wide applicability) that is not achieved by any other known enhancement methods.

In [15], Sun et al. proposed a novel optical transfer function-based micro image enhancement algorithm. An optical transfer function is used by this method for image enhancement. The proposed algorithm was better to save edge information of the micro image and no distortion when it is compared with Butterworth high-pass filter algorithm.

In [16], Ehsani et al. proposed an adaptive and iterative histogram matching (AIHM) algorithm for chromosome contrast enhancement used in medical applications. To meet the different requirements and obtain the different results, some parameters in the presented model could be selected. The detailed simulations were accomplished using different sets of single chromosomes, indicating that the proposed method enhances the details adequately.

In [17], Xiao-guang et al. proposed a generalized fuzzy enhancement method. This method overcomes efficiently the problem of information of middle and low gray levels lost in the method put forward by S.K. Pal. The edge of processed image becomes distinguishable, and by using this method, better object information of low gray levels is kept. It gives highly enhanced image and improves contrast ratio.

In [18], Ke et al. proposed an innovative image enhancement framework consisting of BiTA and SWCE. The traditional gamma adjustment curve is modified using the BiTA algorithm. The adjustment of global luminance is done by BiTA. By analyzing and setting the parameter γ, it limited the contrast lost. The saliency map with a simple contrast enhancement method is integrated by SWCE. SWCE carries out more enhancements in the areas that humans pay larger attention to.

In [19], Tang et al. proposed a new image enhancement technology which is based on a multi-scale contrast measure in the wavelet domain for radiologists to screen the mammograms. It alters the contrast of an image directly, so the proposed image enhancement technology is a direct contrast enhancement technology.

In [20], Dong-liang et al. proposed the generalized fuzzy enhancement method that overcomes the limitations of the traditional fuzzy method of enhancement. The enhancement problem of the low contrast and the narrow gray range images is solved by this method. The improved label algorithm is used for image segmentation and recognition, which is helpful in understanding an image and in object recognition.

In [21], Choi et al. discussed image enhancement technique in which an input RGB color image is transformed into HSV color image. The H component remains unchanged. Firstly, illumination is evaluated using JND-based nonlinear low-pass filter, to enhance the V component. It is done in ratio to the enhanced ratio of V component image, to enhance the S component. Finally, the image is transformed back to RGB color image.

In [22], Hasikin et al. discussed fuzzy gray-scale enhancement technique. The fuzzy measures in the image are maximized by using this technique. By using power law transformation and saturation operator, the membership function is modified. The intensities are increased of the underexposed regions, and intensities are decreased of the overexposed region, which is why the dynamic range is maintained. Other methods use INT operator (the membership value is modified above and below the threshold value) and NINT operator.

In [23], Xianghong et al. discussed the enhancement algorithm for colored medical images. It includes two features: color space transform enhancement and wavelet analysis enhancement. By using wavelet analysis, the image is fragmented and improved. From RGB space, the color image is converted to IHS (intensity, hue, and saturation) space. The new saturation which is enhanced replaces the image saturation. The nonlinear transform adjusts the intensity of the image. The image is then reconverted from IHS space to RGB space, and finally, the image is enhanced.

In [24], Yang et al. discussed morphological reconstruction enhancement. A non-idempotent connected operator is explained in this method. Overall enhancement of simple connected zones of image is done. Finally, the connected zones and the relations of the connected zones of an image are defined which are reconstruction, choice and iteration times. To describe image's peak component, flat area regional maximum, and their relations, a tree structure is used.

In [25], Peng et al. discussed enhancement method of THz images. By using wavelet transform into low and high-frequency part, a THz image is fragmented. Approximation coefficients go through nonlinear enhancement technique. This method magnifies the target and represses the background. Detail coefficients go through histogram equalization but it also enhances noise as well. So firstly, to remove noise in high-frequency band wavelet, denoising is applied. In target identification, the proposed method has good application prospect.

In [26], Gorai et al. proposed an automatic color enhancement technique using particle swarm optimization. Parameters of objective function are made effective by particle swarm optimization. The intensity transformation uses local as well as global information. The objective function makes the use of entropy and edge information for measuring the quality of input image. To obtain the enhanced image, scaling (or sometimes rescaling in case of gamut problem arises) is done.

In [27], Verma et al. used genetic algorithm to enhance an image. GA enhances the image naturally as it measures the fitness of a particular by estimating the intensity of spatial edges comprised in the image. Image is broken into sub images. Transformation function is applied. Fitness function is used. One with higher fitness is selected using tournament selection. Crossover is done using an arithmetic crossover, and finally, PCA-mutation is done.

In [28], Khan et al. proposed fingerprint image enhancement using multi-scale DDFB-based diffusion filters and modified Hong filters. The major goal of a fingerprint image enhancement is to magnify the image in order to remove noise and magnify reliable minutiae points. To calculate orientation field of a fingerprint image, multi-scale DDFB is required. This shows that this technique is more reliable than any derivatives-based techniques.

In [29], Raju et al. proposed a new fuzzy logic and histogram-based algorithm for magnifying low contrast color image. In this method, there are two parameters: K and M, where K is contrast identification parameter and M is image's average intensity parameter. The RGB image is transformed to HSV color space. Under the control of parameter M, V component is strained in order to magnify the image. The visual quality is modified by this method.

Table 2 Analysis of research papers related to various image enhancement techniques

Paper ID	Technique(s) identified
[1], [9]	Laplacian Method (Mask and Pyramid)
[2], [10], [16], [20], [22], [29]	Wavelet transform-based techniques
[3], [8]	Nonlinear image enhancement methods
[13], [26], [30]	Histogram-based techniques
[11], [24]	Genetic algorithm-based enhancement
[14], [17], [19]	Fuzzy-based enhancement
[21], [25], [27]	Multi-scale methods for enhancement
[4], [5], [6], [7], [12], [15], [18], [23], [28]	Other methods (HSV, DoG, PSO, shearlet transform, FPGA, optical method, BiTA and SWCE, FICE, etc.)

In [30], Lin et al. proposed retinex improvement for nighttime image enhancement. In case of low light images, this method can be applied to preserve normal or intensive lighting areas and reduce noise in highly low light areas.

In [31], Samantaray et al. focus on improving fingerprints using a proposed difference of Gaussian method. For that, histogram equalization is used and the proposed method is compared with other approaches also. Separate algorithms for contrast enhancement and detail enhancement are proposed so that fingerprints can be analyzed correctly.

In [32], Chouhan et al. propose a dynamic stochastic resonance-based technique which is used to improve the quality of dark contrast images so that they can be easily identified. For this method to be implemented, image must be in discrete wavelet transform (DWT) domain. For the purpose of contrast enhancement, internal noise of the image is used instead of external noise which improves the performance of enhancement.

In [33], Akila et al. compare various contrast enhancement techniques for images used in breast cancer detection (mammographic image enhancement). Techniques used are HE, CLAHE, BBHE, MMBEBHE, and RMSHE. According to the analysis, HE changes brightness of the image so the details cannot be viewed clearly. BBHE and MMBEBHE provide noise removal from images. Out of CLAHE and RMSHE, which both provide significant contrast enhancement, RMSHE is proved to be better.

4 Conclusion

In this paper, from Sect. 3 and Table 2, research questions proposed earlier in Table 1, i.e., RQ1, RQ2, and RQ3 are identified and answered. The main goal of this systematic review is to identify and analyze various image enhancement techniques which are used to enhance the quality of an image, including noise, contrast, brightness, etc. These techniques help researchers compare various enhancement algorithms and choose the best of them as per the required purpose.

Answer to RQ1

Various techniques that are identified from our review work are listed as follows:

1. **Laplacian method:** mask and pyramid.
2. **Wavelet-based Enhancement:** wavelet transform, lifting wavelet transform, multi-scale contrast feature in wavelet domain, wavelet and color space transform, wavelet-based enhancement using DSR.
3. **Nonlinear Image Enhancement Methods:** singular value decomposition (SVD).
4. **Histogram-based methods:** HE, CLAHE, BBHE, MMBEBHE, and RMSHE, adaptive and iterative histogram matching (AIHM), fuzzy logic and histogram-based image enhancement.
5. **Genetic Algorithm-based evolutionary enhancement** (EVOLEHA).
6. **Fuzzy-based enhancement:** fuzzy-based iterative image enhancement, generalized fuzzy-based image enhancement.
7. **Multi-scale methods:** multi-scale morphological reconstruction, multi-scale retinex improvement, multi-scale DDFB-based diffusion filters and modified Hong filter.
8. **Other techniques:** field programmable gate array, stepwise refinement method, FICE algorithm, discrete shearlet transform, optical transfer function-based image enhancement, BiTA and SWCE, HSV Transform, particle swarm optimization, difference of Gaussian.

Answer to RQ2:

The most common algorithm traditionally used for image enhancement is histogram equalization. Other algorithms were also used but they were somehow an extension or modification of histogram equalization only.

As the technology advanced, various new methods were developed for the enhancement of images. As expected, new techniques were better than the traditional image enhancement techniques. As studied from various papers, we found out that wavelet transform- and soft computing-based methods (GA and fuzzy) were better than the traditional histogram equalization methods.

Answer to RQ3

As analyzed, there are various factors which can affect image quality. Some of them are noise, sharpness, distortion, contrast, color accuracy, dynamic range, exposure accuracy, lens flare, etc. These factors must be kept in mind while choosing or designing any image enhancement algorithm.

Finally, we can conclude that main techniques identified are histogram equalization, wavelet processing and transform, genetic algorithms, fuzzy enhancement techniques, Laplacian and Gaussian techniques, etc.

In future, this review can be used to compare various image enhancement algorithms and their feasibility can also be determined. Also, a specific area can be chosen and further review and research work can be carried out. The proposed studies may be used to do comparative analysis, and further improvement may be applied by proposing new image enhancement techniques. Further, proposed image enhancement techniques may increase the quality of images.

References

1. Singh, P.K., Sangwan, O.P., Sharma, A.: A Systematic Review on Fault Based Mutation Testing Techniques and Tools for Aspect-J Programs, published in 3rd IEEE International Advance Computing Conference, IACC-2013 at AKGEC Ghaziabad, IEEE Xplore, pp. 1455–1461, 2013
2. Singh, P.K., Panda, R.K., Sangwan, O.P.: A Critical Analysis on Software Fault Prediction Techniques, published in World Applied Sciences Journal, Vol. 33, No. 3, pp. 371–379, 2015
3. Singh, P. K., Agarwal, D., Gupta, A.: A Systematic Review on Software Defect Prediction, published in Computing for Sustainable Global Development (INDIACom), IEEE, pp. 1793–97, 2015
4. Negi, S.S., Bhandari, Y.S.: A hybrid approach to Image Enhancement using Contrast Stretching on Image Sharpening and the analysis of various cases arising using histogram, published in Recent Advances and Innovations in Engineering (ICRAIE), pp. 1–6, 2014
5. Wu, C., Liu, Z., Jiang, H.: Catenary image enhancement using wavelet-based contourlet transform with cycle translation, published in Optik-International Journal for Light and Electron Optics, Vol. 125, No. 15, pp. 3922–3925, 2014
6. Wang, L.J., Huang, Y.C.: Non-linear image enhancement using opportunity costs, published in Second International Conference on Computational Intelligence Communication Systems and Networks (CICSyN), IEEE, pp. 256–261, 2010
7. Sree, V.K., Rao, P.S.: Hardware implementation of enhancement of retinal fundus image using simulink, published in IEEE Asia Pacific Conference on Postgraduate Research in Microelectronics and Electronics (PrimeAsia), pp. 239–244, 2013
8. Yaping, L., Jinfang, Z., Fanjiang, X., Xv., S.: The recognition and enhancement of traffic sign for the computer-generated image, published in Fourth International Conference on Digital Home (ICDH), pp. 405–410, 2012
9. Imtiaz, M.S., Khan, T.H., Wahid, K.: New color image enhancement method for endoscopic images, published in International Conference on Advances in Electrical Engineering (ICAEE), IEEE, pp. 263–266, 2013
10. Premkumar, S., Parthasarathi, K.A.: An efficient approach for colour image enhancement using Discrete Shearlet Transform, published in 2nd International Conference on Current Trends in Engineering and Technology (ICCTET), IEEE, pp. 363–366, 2014
11. Bhattacharya, S., Gupta, S., Subramanian, V.K.: Localized image enhancement, published in Twentieth National Conference on Communications (NCC), IEEE, pp. 1–6, 2014
12. Teng, Y., Liu, F., Wu, R.: The Research of Image Detail Enhancement Algorithm with Laplacian Pyramid, published in IEEE International Conference on Green Computing and Communications, pp. 2205–2209, 2013
13. Wang, X.B.: Image enhancement based on lifting wavelet transform, published in 4th International Conference on Computer Science & Education, IEEE, pp. 739–741, 2009
14. Munteanu, C., Rosa, A.: Gray-scale image enhancement as an automatic process driven by evolution, published in IEEE Transactions on Systems, Man, and Cybernetics, Vol. 34, No. 2, pp. 1292–1298, 2004
15. Sun, Y., Yin, X.: Optical transfer function-based micro image enhancement algorithm, published in 2013 IEEE International Conference on Communications Workshops (ICC), pp. 959–963, 2013

16. Ehsani, S.P., Mousavi, H.S., Khalaj, B.H.: Chromosome image contrast enhancement using adaptive, iterative histogram matching, published in 7th Iranian conference on Machine Vision and Image Processing (MVIP), IEEE, pp. 1–5, 2011
17. Xiao-Guang, Z., Ding, G., Jian-Jian, X.U.: Generalized fuzzy enhancement of image for radiographic testing weld, published in Proceedings of the 4th International Symposium on Image and Signal Processing and Analysis, IEEE, pp. 94–99, 2005
18. Ke, W.M., Chen, C.R., Chiu, C.T.: BiTA/SWCE: Image enhancement with bilateral tone adjustment and saliency weighted contrast enhancement, published in IEEE Transactions on Circuits and Systems for Video Technology, Vol. 21, No. 3, pp. 360–364, 2011
19. Tang, J., Sun, Q., Agyepong, K.: An image enhancement algorithm based on a contrast measure in the wavelet domain for screening mammograms, published in IEEE International Conference on Image Processing, Vol. 5, pp. 5–29, IEEE, 2007
20. Dong-liang, P., An-Ke, X.: Degraded image enhancement with applications in robot vision, published in IEEE International Conference on Systems, Man and Cybernetics, Vol. 2, pp. 1837–1842, IEEE, 2005
21. Choi, D.H., Jang, I.H., Kim, M.H., Kim, N.C.: Color image enhancement based on single-scale retinex with a JND-based nonlinear filter, published in IEEE International Symposium on Circuits and Systems, IEEE, pp. 3948–3951, 2007
22. Hasikin, K., Isa, N.A.M.: Enhancement of the low contrast image using fuzzy set theory, published in UKSim 14th International Conference on Computer Modelling and Simulation, IEEE, pp. 371–376, 2012
23. Xianghong, W., Shi, Y., Xinsheng, X.: An effective method to colour medical image enhancement, published in IEEE/ICME International Conference on Complex Medical Engineering, pp. 874–877, IEEE, 2007
24. Yang, S., Wang, C., Deng, L.: A new Approach of Image Enhancement Based on Multi-scale Morphological Reconstruction, published in Ninth International Conference on Hybrid Intelligent Systems, Vol. 1, pp. 113–116, IEEE, 2009
25. Peng, Z., Weiliang, H., Wenjian, L., Zhihui, Z.: The multi-scales nonlinear enhancement method of THz image, published in 2013 IEEE International Conference on Microwave Technology & Computational Electromagnetics (ICMTCE), pp. 341–344, 2013
26. Gorai, A., Ghosh, A.: Hue-preserving color image enhancement using particle swarm optimization, published in Recent Advances in Intelligent Computational Systems (RAICS), pp. 563–568, IEEE, 2011
27. Verma, A., Goel, S., Kumar, N.: Gray level enhancement to emphasize less dynamic region within image using genetic algorithm, published in 3rd International conference on Advance Computing Conference (IACC), pp. 1171–1176. IEEE, 2013
28. Khan, T.M., Khan, M.A., Kong, Y.: Fingerprint image enhancement using multi-scale DDFB based diffusion filters and modified Hong filters, published in Optik-International Journal for Light and Electron Optics Vol. 125, No. 16, pp. 4206–4214, 2014
29. Raju, G., Nair, M.S.: A fast and efficient color image enhancement method based on fuzzy-logic and histogram, published in AEU-International Journal of electronics and communications, Vol. 68, No. 3, pp. 237–243, 2014
30. Lin, H., Shi, Z.: Multi-scale retinex improvement for nighttime image enhancement. published in Optik-International Journal for Light and Electron Optics Vol. 125, No. 24, pp. 7143–7148, 2014
31. Samantaray, S., Bakshi, S., Sa, P.K.: An efficient DoG based fingerprint enhancement scheme, published in Proceedings of the International Conference on Advances in Computing, Communications and Informatics, pp. 547–550. ACM, 2012
32. Chouhan, R., Jha, R.K., Biswas, P.K.: Wavelet-based contrast enhancement of dark images using dynamic stochastic resonance, published in Proceedings of the Eighth Indian Conference on Computer Vision, Graphics and Image Processing, pp. 73, ACM, 2012
33. Akila, K., Jayashree, L.S., Vasuki, A.: Mammographic Image Enhancement Using Indirect Contrast Enhancement Techniques–A Comparative Study, published in Procedia Computer Science Vol. 47, pp. 255–261, 2015

Virtual Field and Wave Laboratory Using Scilab and Java

Sarika Agarwal and Gunjeet Kaur

Abstract Combining the open source tools to efficiently process the data can enable anyone to experiment, explore as well as share their outcomes with the rest of the community. This paper describes and implements Virtual and Remote Laboratory using Scilab and Java. The main objective of this lab is to improve the study of field and wave theory by allow the user to work at any place without installing Scilab and Java. In addition to this, the application also provides Graphical user Interface (GUI) on Web where user can alter the coordinates of amplitude, frequency of wave, direction, wavelength, and time. Thus, students can develop the Scilab program on different parameters and then deploy to see the results in form of graph and Text.

Keywords Wave propagation · Scilab · Java · JavaSci

1 Introduction

For advanced learning techniques, it is essential to use laboratories that enable to practice the acquired knowledge and to monitor the real time errors or issues that can appear in a real system but do not exist in virtual or simulated laboratories [1, 2]. Hence, the Web-based laboratories are an interesting approach. Remote laboratories are E-learning resources provide a distance teaching framework and improve the accessibility of experimental setups which meet the user's hands-on learning needs in this methodology [3–6].

The proposed paper presents virtual laboratory for the scholars of the Field and wave Theory subjects. The idea is based on the distance learning where students can do their laboratory work on Web. In this work, first we discuss about the

S. Agarwal (✉) · G. Kaur
Dronacharya Group of Institutions, Greater Noida, India
e-mail: sarikagarwal.it@gmail.com

G. Kaur
e-mail: gunjeet.18@gmail.com

S. Urooj and J. Virmani (eds.), *Sensors and Image Processing*, Advances in Intelligent Systems and Computing 651, https://doi.org/10.1007/978-981-10-6614-6_24

experiment shown on Web, the parameters that can change by user/student. In II part, the connectivity of Scilab and Java is discussed.

2 Virtual and Remote Laboratory on Web

The Remote and virtual laboratory can be separated into two parts Fig. 1. The first part consists of client side that is developed through Java Server Pages and servlets. The second part is server part that is developed through JavaSci, Scilab, Tomcat Server, and servlet. The protocols used in the server are FTP, HTTP, TCP/IP. Using a stand-alone application students can easily access the laboratories.

Considering the first case, the interface can be executed by Java which can directly exchange the data with the laboratory using three main protocols. File transfer protocol, i.e., FTP is used to load code to the server and keep results, second protocol, i.e., TCP-IP is used to converse with Scilab and HTTP to exhibit on browser which used to communicate and exchange information with the server.

The subsequent part is the server side which is composed by the connectivity of Scilab and Tomcat Server. Server executes the student code and drives these outcomes in form of Graph or text to the user.

This paper describes the application of plotting the plane wave for +x, +y, +z direction for various values of t (Time). The user can change the Amplitude of wave, frequency of wave, Wavelength of wave, and Time and then executes the application. The Scilab program along with input given by user goes to the Tomcat Server then communicates with Scilab to execute the code and send back the result to Tomcat and Tomcat again display the result to the user. The result may be in text, graphics, or both.

Figure 2 shows the Graphical user Interface of virtual laboratory application of study propagation of plane waves. A wave in which the parallel wave fronts are of constant amplitude and are normal to the phase velocity vector, i.e., the direction of propagation is referred to a place wave. Therefore, its magnetic and electric field lie in a plane which is perpendicular to the direction of wave travel. If the field, in any

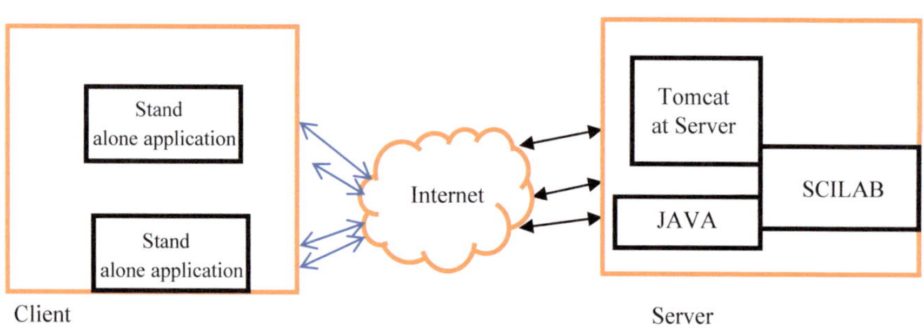

Fig. 1 Components of Remote and virtual laboratory

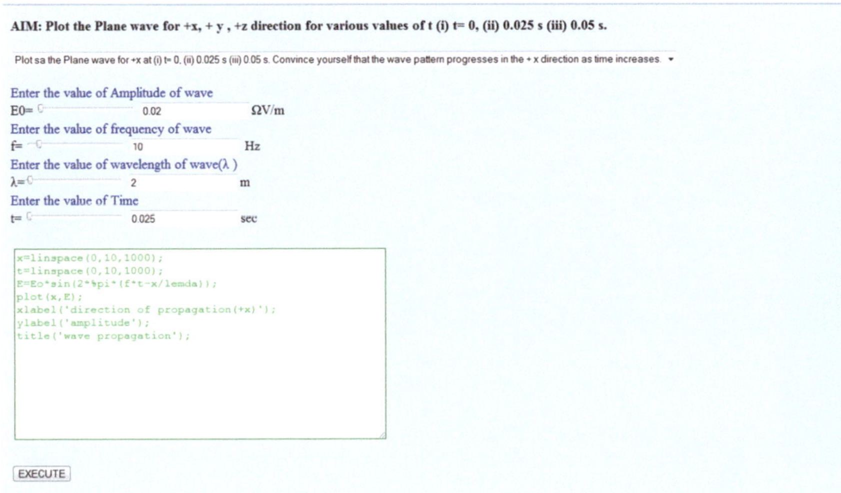

Fig. 2 User view of experiment

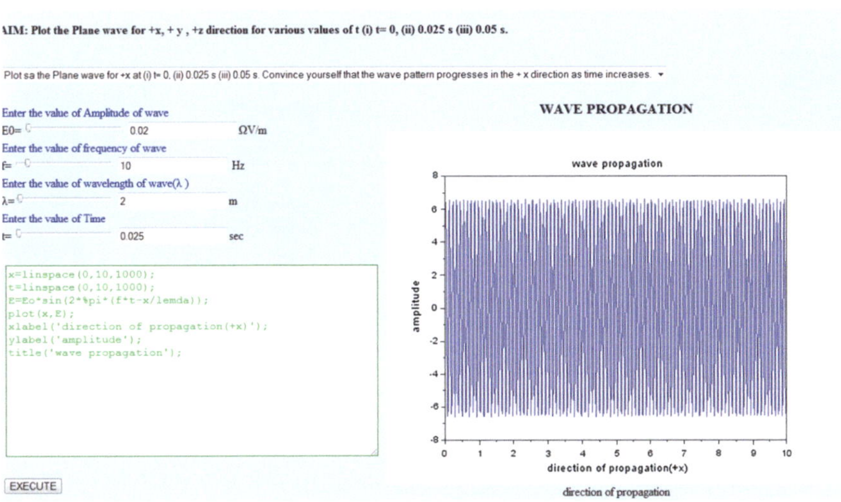

Fig. 3 Remote Laboratory plot a wave for +X Axis

plane, has the same magnitude and direction at each point and is at right angles to the path of wave travel then the wave is referred as Uniform plane wave.

If we consider the case for the propagation of wave in lossless medium, then the conduction current is almost absent in comparison with displacement current (i.e., $\varphi = 0$). For the analysis, we have Helmholtz equation. So far a case of uniform plane wave travelling in x-direction Fig. 3, 4, and 5.

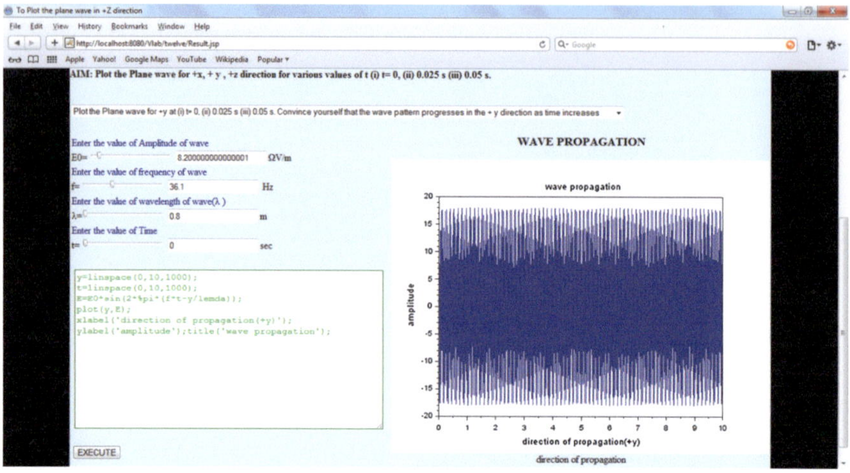

Fig. 4 Remote Laboratory plot a wave for +Y Axis

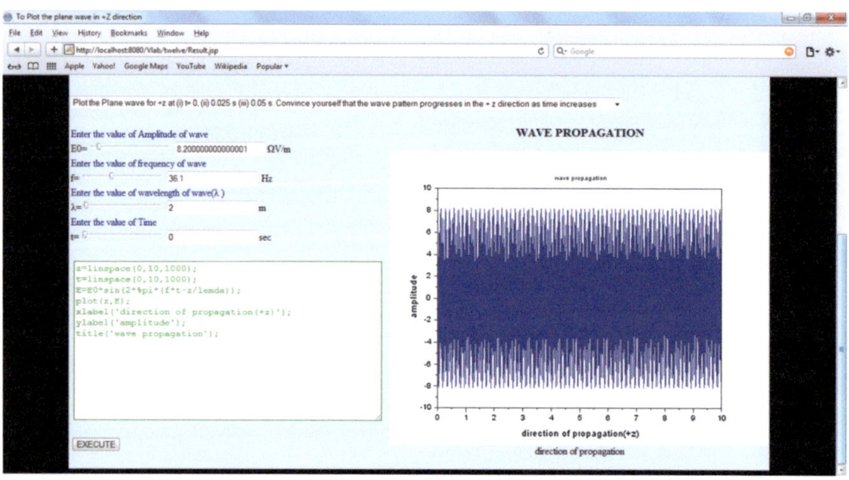

Fig. 5 Remote Laboratory plot a wave for +z Axis

3 Server

The server is composed of three elements, the Scilab that takes program as input and gives graph or text as output, Tomcat Server that communicates between Scilab and browser, Web-component as SERVLET/JSP is a program language used to implement server.

1. Scilab: Scilab is a software which is open source and cost-free for mathematical calculations and computations which provide a very dominant environment for scientific and engineering purposes. This software also consists of hundreds of numerical functions. Advance data structures; 2-D and 3-D graphical functions can be accessed using this software as it has a high level programming.
2. Tomcat Server: Apache Tomcat™ is also an open source software that is an realization of the Java Server Pages, Java Expression Language, Java servlet, and Java Web Socket technologies. It takes the user code, executes it and sends the resulting control output to the user/browser.
3. Connectivity of Scilab with Tomcat through JavaSci: Scilab should offer a Java interface as Java is a key programming language. A documented and simple API is provided by JavaSci which pursues the Java conventions to use Scilab engine features from a Java application or library.

3.1 Connectivity of Scilab and Java

Scilab software is a type of numerical algorithm language and strong matrix operation. One can connect Scilab with JavaSci and with Command Line. We had used both for Text result we had connect Scilab with JavaSci and For Graphic output we had connect Scilab with Java on Command Line.

3.1.1 Connecting Scilab on Command Line with Java

```
< %
Process out = Runtime.getRuntime().exec ("scilex -nw -nb –e m = [1 2 3; 3 4 5];
disp(m);exit;");
%>
```

Nw = No window
Nb = Welcome message should not be displayed
E = an expression in the subsequent item which the Scilab engine must execute.

3.1.2 Connectivity of Scilab with JavaSci in Text Mode and Graphic Mode

JavaSci is a tool for Call Scilab engine from a Java application. To connect Scilab with Java, JavaSci provides two API "*org.scilab.modules.javasci.Scilab*" provides the classes to connect with Scilab with Text or Graphic mode, the other is *org.scilab.modules.types.ScilabType* that helps to get the output in Scilab datatype.

```
<%
Scilabsci = new Scilab(true);
sci.open();
sci.exec("disp(%pi);");
%>
```

4 Offline Work and Evaluation

Students must write a report with the major outcomes obtained by them after the experiments. They can also download the background of the experiment problem and proposed algorithm. They are encouraged to reconstruct themselves in solving the problems.

5 Conclusions and Future Work

As practical sessions are very much required and mandatory for complete understanding of the basic concepts, the remote and virtual laboratories are being used in various Engineering Institutes for five years. Students as well as lab instructors get motivation and contentment in their work by using these virtually created laboratories.

The students can operate these laboratories using a stand-alone application. Moreover, students can also change the values of inputs to get results accordingly. These experiments facilitate an interface that directly exchanges the data with the server of the laboratory and the outcomes are directly saved by the students and sent to their mentors.

References

1. Chaos, D., Chacón, J., Lopez-Orozco, J.A., Dormido, S.: Virtual and Remote Robotic Laboratory Using EJS, MATLAB and LabVIEW. Sensors. **13**(2), 2595–2612 (2013) doi:10.3390/s130202595
2. Gomes, L., Bogosyan, S.: Current trends in remote laboratories. IEEE Trans. Ind. Electron. **56**, 4744–4756 (2009) [Google Scholar]
3. De la Torre, L., Sánchez, J., Dormido, S., Sánchez, J.P., Yuste, M., Carreras, C.: Two web-based laboratories of the FisL@bs network: Hooke's and Snell's laws. Eur. J. Phys. **32**, 571–584 (2012) [Google Scholar]

4. Esquembre, F.: Easy Java Simulations: A software tool to create scientific simulations in Java. Comput. Phys. Commun. **156**, 199–204 (2004) [Google Scholar]
5. Farias, G., Cervin, A., Årzénn, K.E., Dormido, S., Esquembre, F.: Java simulations of embedded control systems. Sensors. **10**, 8585–8603 (2010) [Google Scholar]
6. https://www.scilab.org/scilab/interoperability/calculation_engine/java

Classification and Comparative Study of IRS LISS-III Satellite Images Using RBFN and Decision Tree

Anand Upadhyay, Santosh kumar Singh, Shailesh Kisan Gaikwad and Ashutosh Chandra Mukherjee

Abstract The satellite image classification is categorization of objects available on earth surface using the satellite images. It is one of the very important areas of satellite images processing to understand the changes on the earth surface. So, the main intention behind the following work is to apply the decision trees and radial basis function network-based supervised classification techniques to understand the land cover and land used area in Mumbai. The classifiers are implemented using the MATLAB simulation toolbox. Here the IRS P6 LISS-III satellite image of Mumbai region is used to classify the areas of Mumbai. The different areas of Mumbai region are classified such as the area is covered by water, forest, wetland, and development areas. Decision tree and RBFN classification techniques are trained with the same set of training data and applied on the same set of the testing data set. So, the experimental result of our classifiers shows that the classification accuracy of the RBFN-based satellite image classification is higher than decision tree-based satellite image classification.

Keywords RBFN (Radial basis function network) · Decision tree · IRS (Indian remote sensing) · LISS-III (Linear imaging and self-scanning sensor) · Classification

A. Upadhyay (✉)
Research ScholarAMET University, Maharashtra, India
e-mail: anandhari6@gmail.com

S.k. Singh
Head of Information Technology Department, Maharashtra, India
e-mail: sksingh14@gmail.com

S.K. Gaikwad · A.C. Mukherjee
Post Graduate Students, Maharashtra, India
e-mail: shaileshgaikwad04@gmail.com

A.C. Mukherjee
e-mail: ashutoshmukherjee20@gmail.com

S. Urooj and J. Virmani (eds.), *Sensors and Image Processing*, Advances in Intelligent Systems and Computing 651, https://doi.org/10.1007/978-981-10-6614-6_25

1 Introduction

Classification is an important part of data mining to recognize the pattern in data. The classification is widely used by the people to classify the different objects into different classes. The satellite image is also data in terms of the image, and our job is to automatically recognize the different objects available on the earth surface. The automatic classification of satellite images generates the thematic map of the earth surface. The satellite image is nothing but the image of earth surface in either multispectral of hyper spectral image format. The different bands of hyper spectral and multispectral image of satellite images consist of the images of different wavelength. The different bands of satellite image consist of specific properties; therefore, the different bands consist of very specific and relative information of the different objects of the earth surface. There are some bands which consist of very good information about water, some are good for vegetation and some are good and very relevant for the forest, soil moisture, etc. So in the classification of satellite image, the classification algorithm categorizes the different objects such as water body, forest, wetlands, soil moisture, and vegetation based on the features of the object which are unique for the object which is available in the satellite image. The object available on the earth surface has the same electromagnetic radiation, so the images of such objects are captured by the satellite camera which is far away from the earth. Therefore in satellite image classification, the training data are collected from one place which is approachable to the person for survey and data collection for ground truth of the object; with the help of ground truth and survey, the different features of the object are extracted from the image and the classifiers are trained with the training data so that it automatically classify the objects in the satellite image. The classification is also of two types which are supervised and unsupervised. When the classification is supervised then the classifiers require the training data, otherwise, in unsupervised classification which is known as clustering the image is classified based on the different similarity patterns among the object.

The following research paper is organized in order to give information about literature review, decision tree, radial basis function network, methodology, results, conclusion, and ends with future enhancement.

2 Literature Survey and Review

Xiaoli Tao et al. presented the work based on unsupervised classification using the techniques of normalized radial basis function neural network. The result of normalized radial basis function is also compared with K-mean clustering. Here the author suggested that the performance of the new neural network model is very much accurate and robust than normal radial basis function. The specified classifier was used on the multispectral satellite image of New England which is acquired by Landsat 7ETM+sensors. [1] Yogesh V Kene et al. had implemented the object-based image classification. Here the radial basis neural network model is used for the satellite image classification. The classifier is trained with the segmented value of the objects in the satellite image. There are different types of segmentation algorithm and techniques available, but here the authors had used the curvelet transformation to extract the segmentation of the image and the authors got the good result with the curvelet transformation-based satellite image classification. [2] H.R. Keshtkara et al. had proposed the method of IRS LISS-III satellite image of northeast Iran using the maximum likelihood and decision tree-based classification techniques. Here the authors got more accuracy for the classification of satellite image using decision compared to the maximum likelihood classifier. [3] Richa Sharma et al. had attempted to implement the decision tree classification technique for the classification of the Landsat TM satellite image data using the open-source software known as WEKA. The result of the decision tree classification is also compared with the ISODATA clustering and maximum likelihood classifier algorithm. Here the decision tree-based classification shows better result compared to the other classifiers. [4] M. K. Ghose et al. had attempted to implement the decision trees classification to classify the LISS-III satellite image. Here the concept of separability matrix concept is used for the different bands of satellite image. The result of the decision-based classification technique is compared with the maximum likelihood classifiers where the performance of decision tree was more than the maximum likelihood classifier [5].

3 Study Area and Characteristics

Mumbai is the capital of Maharashtra which is also known as Bombay. Mumbai is also known as the business hub of India. Here for the classification purpose LISS-III satellite image of Mumbai region is used. The LISS-III satellite image is provided by the IRS-P6 Indian satellite; therefore, it is also called as Resourcesat-1. The IRS-P6 provides LISS-III as well as other types of satellite images. LISS-III satellite image is multispectral satellite image available in four different bands of images. The four different bands of the LISS-III satellite images are 0.52–0.59 microns, 0.62–0.68 microns, 0.77–0.86 microns, and 1.55–1.70 microns of images, respectively. LISS-III satellite image is provided by NRSA, Hyderabad, Telangana 500042, India.

4 Radial Basis Function Neural Network

The Radial basis function network is one types of the artificial neural network. It is based on the concept of the theory of function approximation. It is a multilayer perceptron model where the activation function is used as the radial basis function approximation. The architecture of radial basis function network consists of

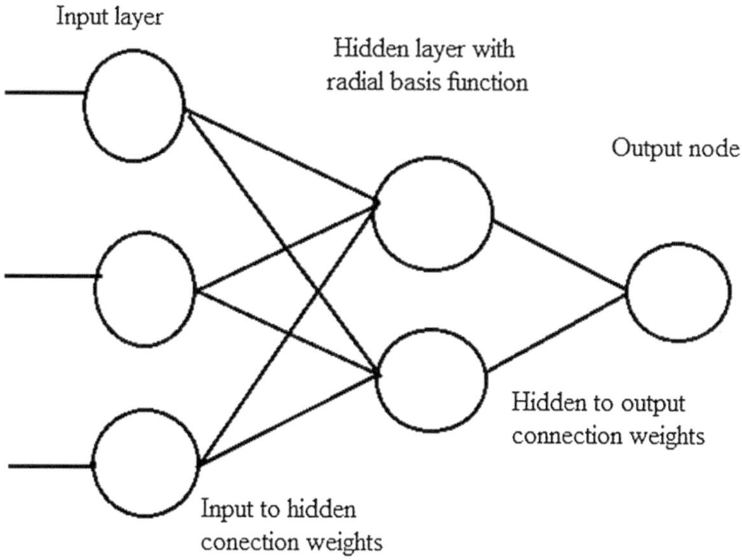

Fig. 1 Architecture of radial basis function network

two-layered feed forward network model. Here each hidden node of radial basis function network consists of the radial basis function. The inputs to hidden layer nodes are connected with weights and similarly the hidden to output nodes are also connected with the weights. The output node applies the linear summation and the value is approximated based on the target assigned to each and every class. The training algorithm in radial basis function network adjusted the weights to get the maximum approximated output at the output node (Fig. 1).

5 Decision Tree

Decision tree is one of the supervised classification techniques for classification which is based on the inductive learning method. The decision tree classification method is very much fast and appropriate for the large amount of the data set. The decision tree classification technique is based on the divide and conquers techniques, so it divides the data exactly in half of the portion, and this is the recursive process until and unless the condition is not achieved. The decision tree algorithm starts with the parent node, and every parent node has their subsequent child nodes. Every child node acts as the parent node for the others node. Each node in the decision tree has the specific decision rule (decision) and based on that further nodes are generated. The decision tree can be applied in many application of classification where speed, dimensionality, and computation matters a lot. It is also capable to handle the missing and noisy data in the classification [6].

6 Proposed Methodology

The methodology is the actual implementation of propose algorithm where the algorithm is implemented and tested on the data, and the performance and result of algorithm are suggested. So, the proposed algorithms are implemented using MATLAB R2010 simulation toolbox. MATLAB is the mathematical computational simulation toolbox which consists of the simulation toolbox for different applications which are from different domains.

The decision tree and radial basis function neural network-based supervised classification methodology are as follows:

6.1 Pre-Processing and Feature Extraction

The important step in satellite image classification starts with the pre-processing of the data. The satellite image consists of different types of noises in the image, so we need to apply different types of filters to remove the noise from the satellite image. After applying the filters on different bands of the images, different bands of LISS-III satellite images are stacked together and one single RGB image is formed. Once the RGB image is formed then based on the survey and ground truth, the feature vectors are collected for the different classes, i.e., water, forest, mangroves, and development. Based on the feature vector of different classes, the training file is prepared, and with the help of training data set, the decision tree and radial basis function neural network are trained.

6.2 Design of Decision Tree and RBFN Classifier

The MATLAB simulation toolbox provides the simulation algorithm of the decision tree and radial basis function network. The feature data set is divided into training, testing, and validation data set which requires for the classifier. The decision tree and radial basis function network expect the feature and class of the different land cover areas along the row of the training data set. Once the decision tree and RBFN are designed then it is trained with the training data set. Now, the decision tree and radial basis function are ready to test and apply for the image classification.

6.3 Testing and Classification

Now, after design and training with the training data set, the decision tree and radial basis function network are ready to test with the testing data set. So based on the testing data set the accuracy of the decision tree and radial basis function neural network is calculated. The confusion matrix and Kappa coefficient are two important techniques which are used to calculate the accuracy of the classifier and make the quantity analysis for the comparative study between the decision tree and radial basis function classifiers. The same trained classifiers are also used to classify the satellite images under the different land cover and land used areas.

7 Result

The Kappa coefficient was actually developed by Cohen in 1960. It is used to measure the observed agreement for categorical data (Landis and Kock 1977). A confusion matrix is also a tool to summarize accuracy of the classifiers. It is basically used for the supervised classification of the data. It consists of the information about classified versus misclassified data in the supervised learning. The diagonal elements in the confusion matrix show the correctly classified data, whereas the non-diagonal elements are misclassified data of the classification. So with the help of that, the accuracy of the classifier is easily calculated, and it will help the algorithm designer to understand the importance and applicability of particular classifier for the specific data.

The below Table 1 and Table 2 show the confusion matrix which is generated after applying the decision tree and radial basis function network on the testing data, respectively.

$$\begin{aligned} \text{Accuracy} &= ((102 + 221 + 67 + 219)/639) * 100 \\ &= 95.31\% \end{aligned}$$

$$\widehat{K} = \frac{N \sum_{i=1}^{r} x_{ii} - \sum_{i=1}^{r} (x_i + * x + i)}{N^2 - \sum_{i=1}^{r} (X_i + * x + i)}$$

K = 0.9338

$$\begin{aligned} \text{Accuracy} &= ((114 + 221 + 77 + 226)/639)*100 \\ &= 99.84\% \end{aligned}$$

Table 1 Confusion matrix of Mumbai region LISS-III by decision tree

Class	Forest	Water	Mangroves	Urban	Total	Users accuracy
Forest	102	0	1	0	103	99.03
Water	8	221	5	0	234	94.44
Mangroves	4	0	67	7	78	85.89
Urban	0	0	5	219	224	97.77
Total	114	221	78	226		
Producers accuracy	89.47	100	85.90	96.90		95.31%

Table 2 Confusion matrix of Mumbai region LISS-III by RBFN

Class	Forest	Water	Mangroves	Urban	Total	Users accuracy
Forest	114	0	0	0	114	100
Water	0	221	0	0	221	100
Mangroves	0	0	77	0	77	100
Urban	0	0	1	226	227	99.56
Total	114	221	78	226		
Producers accuracy	100	100	98.72	100		99.84%

k = 0.9978

The confusion matrix-based accuracy assessment of LISS-III satellite image shows that the accuracy of classification using radial basis function network is more which is 99.84%, whereas the classification accuracy using the decision tree is 95.31%. Similarly, kappa coefficient of the radial basis function network-based classification is also more, that is, 0.9978 which is more that decision-based classification, that is, 0.9338 (Fig. 2, 3, 4, and 5).

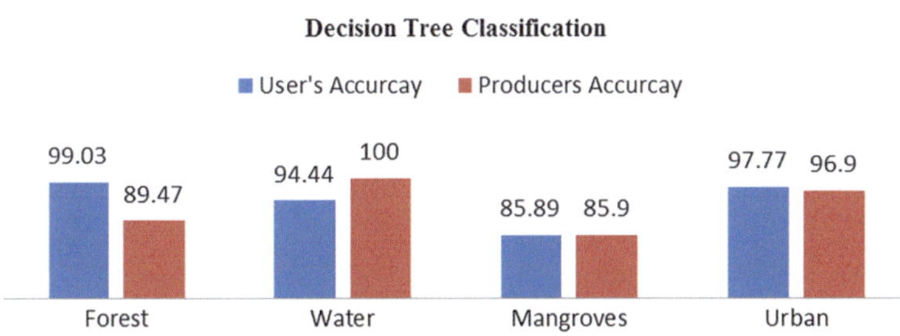

Fig. 2 Classification accuracy assessment of decision tree classification

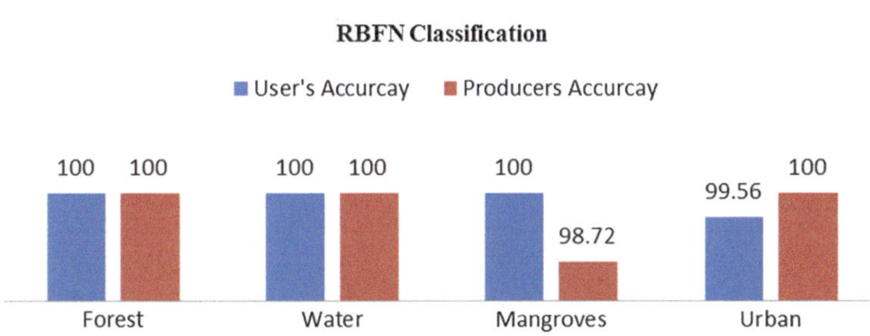

Fig. 3 Classification accuracy assessment of RBFN classification

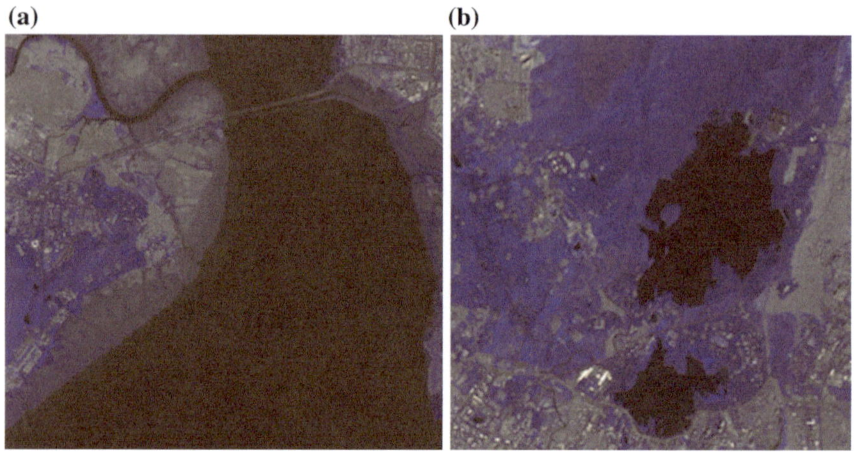

Fig. 4 **a** and **b** LISS-III satellite image before classification

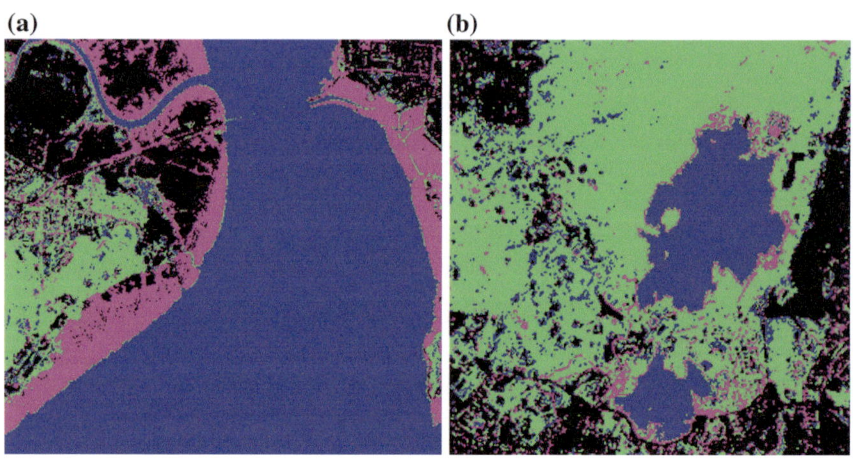

Fig. 5 **a** Colored image after decision tree-based classification **b** Colored image after RBFN-based classification

8 Conclusion

There are the several application of the satellite image classification; therefore, the number of researcher, scientist, and academicians working in the area of satellite image processing suggested the number of the satellite image classifications and performed the comparative study also, but each and every classification algorithms have their advantages and disadvantages so still there are lots of study work and comparative analysis required to understand the classification algorithm for the satellite image. So, in that area, this paper also contributed some efforts where the decision tree and radial basis function network-based classifiers are implemented using the MATLAB Simulink, and it was got from the certain results that the RBFN has good classification accuracy compared to the decision tree classification techniques. The classification accuracy of decision tree-based classification is 95.31%, and radial basis function network based is 99.84%.

Acknowledgements Thanks to NRSA, Hyderabad, Telangana, India for providing the LISS-III data sets free online.

References

1. Tao, X., Michel, H.E.: Classification of multispectral satellite image data using improved NRBF neural networks, Proc. SPIE 5267, Intelligent Robots and Computer Vision XXI: Algorithms, Techniques and Active Vision, 311, October, 2003 doi:10.1117/12.518551
2. Kene, Y.V., Wadkar S.: Object Based Image Analysis of High Resolution Satellite Image using Radial Basis Function Neural Network and Curvelet Transform. European J. Adv. Eng. Techno. **2**(5), 103–107 (2015). ISSN: 2394-658X
3. Keshtkar, H.R., Azarnivand, H., Arzani, H., Alavipanah, S.K., Mellati, F.: Land Cover Classification Using IRS-1D Data and a Decision Tree Classifier. Desert **17**(2), 137–146 (2013)
4. Sharma, R., Aniruddha, G., Joshi, P.K.: Decision tree approach for classification of remotely sensed satellite data using open source support. J. Earth Syst. Sci. **122**(5), 1237–1247 (2013)
5. Ghose, M.K., Pradhan, R., Ghose, S.S.: Decision Tree Classification of Remotely Sensed Satellite Data using Spectral Separability Matrix. Int. J. Adv. Comp. Sci. App. **1**(5), 93–101 (2010)
6. Havyas, V.B., Choodarathnakara, A.L., Thribhuvan, R., Chethan, K.S.: Decision Tree Approach for Classification of Satellite Imagery. Int. J. Soft Computing Eng. (IJSCE) **5**(2), (2015), ISSN: 2231–2307
7. Upadhyay, A., Singh, S.K.: Classification of IRS LISS-III images using PNN, Proceeding of the 9th INDIACom;2nd International conference on computing for sustainable global development, Organized by BVICAM, New Delhi, 11th–13th March, 2015, published in IEEEXplore, ISBN:978-93-80544-15-1

An Analytic Review on Image Enhancement Techniques Based on Soft Computing Approach

Gagandeep Kaur, Nishant Bhardwaj and Pradeep Kumar Singh

Abstract This paper discusses various image enhancement techniques using soft computing approaches. The approaches used are genetic algorithm, fuzzy-based enhancement, neural networks, and optimization techniques (ant colony, bee colony, particle swarm optimization, etc.). The main objective of this paper is to identify the status of currently used soft computing approaches in image enhancement. Our study may help future researchers to overcome the current issues with existing approaches to improve the overall image quality.

Keywords Image enhancement · Image quality · Soft computing approaches · Digital image processing

1 Introduction

Image enhancement is the way of refining the quality of a digital image; thus, making processing of the image easier. After enhancement, the resulting image is better in quality, contrast and can be analyzed correctly by any of the image processing methods. Various techniques can be used for the same, but our study is limited to the techniques based on soft computing approaches. Image enhancement based on soft computing approaches discussed in this paper is genetic algorithm, fuzzy-based enhancement, neural networks, and optimization techniques (ant colony, bee colony, particle swarm optimization, etc.). It has been identified that these techniques provide better enhancement as compared to other traditional image

G. Kaur (✉) · N. Bhardwaj · P.K. Singh
Department of CSE, ASET, Amity University Uttar Pradesh, Noida, India
e-mail: ggndp262@gmail.com

N. Bhardwaj
e-mail: nishant.bhardwaj92@live.com

P.K. Singh
e-mail: pksingh16@amity.edu

S. Urooj and J. Virmani (eds.), *Sensors and Image Processing*, Advances in Intelligent Systems and Computing 651, https://doi.org/10.1007/978-981-10-6614-6_26

enhancement techniques. In addition to this, they are easy to understand and implement.

This paper is divided into four sections. Section 1 is introduction about the image enhancement followed by review procedure in Sect. 2. Section 3 covers the detailed analysis of the selected papers. Finally, conclusions as well as answers to the research questions are discussed in Sect. 4.

2 Review Procedure

To study various image enhancement techniques based on soft computing approaches, we downloaded around 200 papers from IEEE, science direct, ACM, and several other popular digital libraries. We have adopted a similar approach taken by Singh et al. [1, 2] and then identified the most suitable papers using the following steps: (i) we downloaded around 200 papers using standard keywords, (ii) after reading the abstract, title, and conclusion of the downloaded papers, we filtered out 90 relevant papers, (iii) out of the 90 papers selected, we chose 45 most suitable papers related to the suitability of our study, (iv) after that, these papers were thoroughly studied, and the final outcome of each paper was recorded, (v) finally, we have summarized the findings in a tabular format and answered the research questions prepared for conducting systematic review. Following are the research goals to study the image enhancement techniques (Table 1 and 2):

3 Related Work

In this section, various papers related to image enhancement based on soft computing approaches have been taken into account for review and analysis.

In [3], Archana et al. discussed an image enhancement technique using genetic algorithm (GA) which modifies an image with natural contrast. The aim of local contrast enhancement is done by using a, b, c, k parameters with new addition in the range. On the basis of light and dark edges, the gray level of an image is increased. By checking the intensity of spatial edges, the GA measures the individual's fitness. In [4], Deborah et al. proposed a GA technique for magnifying and restoring the old

Table 1 Research questions identified

Research Questions
RQ1: To identify the various techniques used in the image enhancement based on soft computing approach?
RQ2: To compare the traditional image enhancement techniques with the soft computing-based image enhancement techniques?
RQ3: To identify strength and weaknesses of existing image enhancement techniques

Table 2 Search result details

S No.	Soft computing approach used	No. of papers downloaded	No of relevant papers found	Final no of papers chosen for review
1.	Genetic algorithm	42	20	8
2.	Fuzzy enhancement	71	25	18
3.	Neural networks	44	20	11
4.	Optimization technique (PSO, ABC, ACO)	30	10	5
5.	Hybrid	13	10	3
TOTAL		200	85	45

documented image. In this method, the material presents in old documents is retained into digital form. The number of edges having one pixel is the most excellent fitness function. Individuals that have fitness value identical to the fitness value of original image are the optimal solution. In [5], Ueda et al. analyzed an image enhancement system based on GA by acknowledging the liking of the user. GA finds the optimal parameters contained in the selective un-sharp masking of the sharpening system. Based on the liking of the user, the final sharpened image is built. In [6], Radlak et al. considered a method for attaining the optimization visuals by increasing the distance between various colors. The proposed method gives the best visualization result. Using the GA, the final output gives the benefits of pseudo-colorization. In [7], Wu et al. mentioned a method which merges the rough set technique and adaptive genetic algorithm for enhancing an image. Firstly, the image is considered as the knowledge system, and to categorize an image, classification method of rough set theory is used to merge noise attributes and image color attributes. Then, various image enhancement algorithms are applied to modify the image. In [8], Dongzhou et al. analyzed a hybrid genetic algorithm for enhancing an image. It merges the differential evolution with genetic algorithm. For searches and mutation, the fast searching ability of the hybrid genetic algorithm is used. At last, in their work, they have applied transformation function for image enhancement. In [9], Munteanu et al. analyzed an image enhancement method with user behavioral model. In this method, various genetic algorithms are used. This method tries to slightly automate the process of human interpretation which results in rich quality solutions, and hence less human effort is needed. In [10], Daniel et al. proposed a contrast enhancement technique purposely for retinal images. The proposed technique uses masking technique using the green plane and then applying an enhanced GA for the enhancement. The proposed technique is first applied on normal images and then successfully on retinal images. Finally, performance of the proposed technique is evaluated and is found to be better than other traditional techniques.

In [11], Hasikin et al. analyzed a fuzzy set theory method for enhancing the lower contrast image. This method maximizes the fuzzy measurements. The modification of the membership function is done by using the power law transforms and the saturation operator. In [12], Chaira et al. discussed a type II fuzzy enhancement algorithm for image enhancement. Here, Hamacher T Norm is used for defining the new membership functions. It is then compared with other fuzzy image enhancement methods and is observed that the proposed technique gives better contrast images. In [13], Cepeda-Negrete et al. analyzed a fuzzy rule-based system which automatically chooses color constancy algorithms from white patch, gray world, and gray edge for enhancing the color of dark images. Twelve features, including seven colors, three textures, and two lightning features of each image are extracted for processing. In the end, best of the three algorithms is selected for any test image. In [14], Binaee et al. mentioned a noise reducing fuzzy filter which is applied on ultrasound images. It is used for enhancing the quality of ultrasound images, so that they can be efficiently analyzed without any error. The proposed technique contains two steps: detection and filtering. In detection, details of the image are extracted using two fuzzy rule-based system as well as local gradients. In filtering, de-noising of each region is done by a non-local means scheme in which fuzzy similarity is used for finding similarity in weights. In [15], Bing et al. proposed a new algorithm for image enhancement using information fusion for preserving edge information. It is actually a hybrid filter consisting of fuzzy-based region entropy and triangle module fusion operator for edge detection. In [16], Xiao-guang et al. discussed a generalized fuzzy enhancement method. This method overcomes, efficiently, the problem of information of middle- and low-gray levels lost in the method put forward by a previous study. The edge of processed image becomes distinguishable, and by using this method, better object information of low gray levels is kept. It gives highly enhanced image and improves contrast ratio. In [17], Balti et al. implemented fuzzy C-means algorithm for fingerprint image segmentation. Firstly, statistical features of the fingerprint image are extracted and then are used in the fuzzy C-means algorithm. K means algorithm is also used for comparison with the fuzzy C-means algorithm. Also, enhancement of the fingerprint is done by first calculating mean and variance of an image and then finding difference between the original and mean, as well as original and variance images. Resulting images are enhanced than the original image. In [18], Chaira et al. proposed a new technique for medical image enhancement. The technique consists of a Sugeno fuzzy system which uses an intuition-based method for finding membership values. This method is applied on the medical image to convert it into fuzzy image, and then the parameters are optimized using intuitionistic fuzzy entropy. In [19], Zhang et al. mentioned an improved edge detection and image enhancement algorithm which overcomes the shortcomings of algorithms proposed by a previous study for the same. Instead of choosing random threshold parameter value for enhancement, the proposed new method uses OTSU threshold which appropriately chooses the enhancement parameter and can be adjusted according to the requirement. The proposed method simplifies the complex transformation calculation of fuzzy space values and hence improves the performance of the edge detection and image enhancement process. In [20], Wang et al. proposed a new image enhancement algorithm which is an improvement of the Pal–King algorithm. It implements a new membership function which is used to map the image from space domain to fuzzy domain. Nonlinear transformation is then repeatedly applied on the image for enhancement. Finally, inverse operation is applied to get enhanced gray scale values in spatial domain. In [21], Xian-Jiu et al. analyzed a new image enhancement technique for algae microscopic images. It is based on subdividing the image into low-pass part and high-pass part using filtering, enhancing each segment and combining the segments using fuzzification technique to get enhanced image. In [22], Wu et al. mentioned an algorithm for contrast enhancement and detection of edges of a blurred image. It is presented to extract as much information as possible from a blurred image. Algorithm used for the same is a multilevel fuzzy edge detection algorithm which is computationally fast from other algorithms and uses fuzzy operators for enhancement and edge detection. In [23], Jia et al. proposed a computationally fast and efficient fuzzy-based image enhancement algorithm. It uses a new linear membership function for converting the image from space domain to fuzzy domain. Then, for enhancement, the generalized fuzzy operator is applied. The new method shows improvement over the traditional image enhancement algorithms. In [24], Jinping et al. discussed a new improved fuzzy enhancement and edge detection algorithm which uses a Gaussian function. Gaussian function provides a notable improvement in the image's quality and edge extraction. In [25], Jaya et al. discussed a fuzzy rule for magnifying the image having lower contrast in the transformation domain. On both medical and general images, the test is carried out. The final output is compared with fuzzy INT operator. To assess the visualization quality and contrast of an image SDME, IEM, and EME metrics are properly used. In [26], Saeed et al. proposed an image's contrast magnification using the fuzzy set theory. In this method, the image is magnified by using the fuzzy set. The proposed algorithm is analyzed with all other algorithms. This algorithm is more preferable in terms of speed with respect to other algorithms. In [27], Raju et al. proposed an algorithm by using the fuzzy logic and histogram for magnifying an image. There are two parameters which are used in this method. The first parameter is K which is contrast identifying parameter, and the other one is M which is average intensity parameter. The RGB image is transformed to hue, saturation, and value (HSV) image. The strain is put on V component under the control of M parameter, so that the image can be magnified. In [28], Hanmandlu et al. implemented an image enhancement method by using the fuzzy intensification. In this method, the hue, saturation, and value (HSV) model's saturation and histogram are converted into the domain of fuzzy. The NINT operator is used in this method. The image's intensity specifications and the fuzzifier are assessed by effectively using the entropy and contrast in the domain of the fuzzy. This method is applicable for the images having the lower contrast and lower intensification.

In [29], Alilou et al. discussed a general regression based neural network technique which is used for image restoration and inpainting. In this technique, the missing regions of the image are defined by applying the GRNN to perform regression over the neighbor pixels. The proposed technique was found to be computationally efficient and easy to implement. It was tested on various gray scale and color images. In [30], Chitwong et al. discussed different image enhancement techniques by applying the competitive Hopfield neural network. The precision of clustering is better in simulated image and is shown by competitive Hopfield neural network. The enhancement's performance is also finer than he fuzzy C-mean in different clusters for an image. In [31], Nieuwenhuis et al. proposed an adaptive filtering concept. It provides a framework to compose aforementioned high-level knowledge and using it to fix the filtering parameters locally. The performance is improved by using binary rotation of data and multiple class principal component analysis (PCA). In [32], Yin et al. discussed an ultrasound image modifying technique by applying the neural network. Mapping is formed by the artificially neural network. To the artificially neural network, a parameter is computed. The fusion of data is done by the fuzzy logic. To filter the sample data, an algorithm of local histogram is used. The algorithm that is proposed in this paper effectively maps with high-resolution image and low-resolution image. In [33], Zhang et al. considered a technique for high resolution by applying Hopfield neural network. In this method, the Hopfield neural network solves the ill-positioning super resolution problem. This method considers both the blurring and noise in an image and creates a great resolution image with no noise. The image produced by this method has best quality in respect of peak signal-to-noise ratio (PSNR). In [34], Pan et al. proposed a method for diagnosing different types of plants by digital images of leaves. In this method, the clear objects are made in the source image by the image enhancement process. After this, the image blocks are apportioned from the source because of different kinds of size and shape of the block of the image. By applying the image analysis tools, the properties like radius, sizes, solidity, eccentricity, and perimeters are calculated using neural network approach. In [35], Singh et al. analyzed an image enhancement technique for screening the luggage. In the neural network, the input is viewability measures and the output is the image enhancement algorithm. The best choice of the image enhancement algorithm is done by the prediction systems. Then, the prediction system and baseline system approach are compared. In [36], Ma et al. mentioned an algorithm for enhancing the image by using the pulse-coupled neural network and rough set. Firstly, the noise of the image is detected using pulse-coupled neural network time matrix and divides the image in three parts of image based on the noise features and intensity features. After this, denoise the image by applying filter method. At last, enhance the different parts of image by applying the various methods. This proposed algorithm shows better results for the image in the darker regions, and noise is also decreased. In [37], Rao et al. discussed the different types of image enhancement techniques. The good results are shown by the classical spatial filters, and there is very little complexity for computation. A multilayer self-organizing neural network with error back propagation depends on descent gradient is compared. A grand requirement of computations is required because of the adjustment of weights in every layer. In each iteration, the error is negligible. The computation speed of cellular neural network is very high when it's related to neural network. In [38], Weixin et al. implemented a X-ray image enhancement technique by using the Hopfield neural network. In this method, by using a suitable energy function, the image enhancement problem is converted to the optimizing problem. The procedure of normalization is not used in this method because the function for evaluating the quality is used for optimizing the image. In [39], Varghahan et al. discussed a technique for enhancing the handwritten data by applying the neural network and threshold technic*al*. In this paper, various filtering techniques are discussed. For cleansing the image and reducing the noise, artificial neural network (ANN) showed the best result. Cleansing image result along with threshold technical is used for the training of the multilayer perceptron network.

In [40], Shanmugavadivu et al. discussed a contrast enhancement technique using the basic concept of histogram equalization. In this proposed technique, the image histogram is first divided into two parts using Otsu's threshold. After that, a set of optimized weighing constraints, which are formulated using particle swarm optimization (PSO), are applied over the two parts. The two parts are then equalized independently and then are combined to produce a contrast enhanced and brightness preserved image. In [41], Draa et al. implemented a meta-heuristic search technique named artificial bee colony (ABC) algorithm for image enhancement. In this, old gray levels of the input image are replaced by new enhanced gray levels. These new gray levels are searched for using the ABC algorithm. It is then compared with traditional image enhancement approaches as well as GA and is found to be better almost all of them. In [42], Gorai et al. discussed a particle swarm optimization (PSO)-based image enhancement technique. PSO depends on objective function optimized the parameters. The local information and global information of an input image are used by the intensity transformation function. To measure the quality of an image, the edge information and entropy are acknowledged by the objective function. With the help of scaling, the enhanced image is achieved. In [43], Benala et al. proposed an ABC optimizing algorithm for magnifying the image. This algorithm provides excellent magnitudes for an image. It is better than GA. The chance of dropping into local optimum is less in ABC algorithm. The main goal of this algorithm is to advance the local search ability of GA by maintaining the global search ability. In [44], Hanumantharaju et al. discussed a technique for magnifying an image by applying the particle swarm optimization. The algorithm uses criteria based on the information of edge and entropy of the image. By using this method, the specifications of the multiscale retinex like number of scales, Gaussian surround space constants, etc., are optimized. Their proposed algorithm resulted better image enhancement as compared to existing algorithms like histogram equalization, filtering, etc.

In [45], Zhou et al. explored a fuzzy-based image enhancement algorithm whose parameters are optimized using GA. In the proposed technique, firstly image is transformed from spatial domain to fuzzy domain. Then, membership function is optimized using GA and applied on the fuzzy image. Finally, image is converted back to space domain using de-fuzzification and enhanced image is obtained. Further, enhancement can be done by applying the same process again. In [46], Zhang et al. discussed a satellite cloud enhancing technique. In this method, the nonlinear transform parameters are adjusted to modify satellite cloud images. This method gives a reference for location of center and the prediction of the typhoon's intensity. In [47], Hoseini et al. considered a new hybrid technique for image enhancement. It combines the evolutionary techniques of GA, ant colony optimization (ACO), and simulated annealing (SA). Improvement is done on the fitness

Table 3 Various techniques Identified

Technique identified	Study ID
Genetic algorithm	[3–10]
Fuzzy-based enhancement	[11–28]
Neural network-based enhancement	[29–39]
Optimization algorithms	[40–44]
Hybrid approach	[45–47]

function of GA, whereas the ACO and SA techniques remain same. The combined algorithm is then compared with other traditional image enhancement techniques like linear contrast stretching, histogram equalization, fuzzy, etc. and is found to be better.

After reviewing all these 45 research papers, we have identified mainly five soft computing techniques as mentioned in Table 3 along with paper ID.

4 Conclusions

Research questions RQ1, RQ2, and RQ3 taken in Sect. 1 for systematic review are put forward in Table 1 and resolved after doing the analysis of papers in Sect. 3.

Answer to RQ1: various soft computing approaches used in this paper for image enhancement are GA, fuzzy-based enhancement, neural networks, and various optimization techniques like ABC, ACO, PSO, and hybrid approaches.

Answer to RQ2: the traditional algorithms used for image enhancement like histogram equalization, wavelet transform-based enhancement, contrast adjustment, etc., are compared with the enhancement techniques based on soft computing approaches. It is found after analyzing various papers that the soft computing approach is better than the traditional approaches since performance is better, which is shown in various papers. Better image enhancement results may be achieved using hybrid soft computing approaches, meta-heuristic approaches.

Answer to RQ3: strengths of existing image enhancement techniques

1. They can be used for linear stretching and visual perception when image has close contrast data.
2. They remove noise from the images and provide sharp image by applying local enhancement.
3. Histogram equalization is the most popular enhancement technique which calculates intensity of pixels and uses them for enhancement.

Weaknesses of existing image enhancement techniques

1. Existing image enhancement algorithms are very computation intensive and require a large amount of memory to store the intermediate data.

2. Algorithms are quite complex to understand and implement.
3. Very few techniques are practically used for image enhancement which leaves a large scope for new enhancement techniques.

References

1. Singh, P.K., Sangwan, O.P., Sharma, A.: A Systematic Review on Fault Based Mutation Testing Techniques and Tools for Aspect-J Programs, published in 3rd IEEE International Advance Computing Conference, IACC-2013 at AKGEC Ghaziabad, India, IEEE Xplore, pp. 1455–1461, 22–23, February 2013
2. Singh, P.K., Agarwal, D., Gupta, A.: A Systematic Review on Software Defect Prediction, published in Computing for Sustainable Global Development (INDIACom), IEEE, pp. 1793–97, 2015
3. Verma, A., Goel, S., Kumar, N.: Gray level enhancement to emphasize less dynamic region within image using genetic algorithm, published in 3rd International Advance Computing Conference (IACC), pp. 1171–1176, IEEE, 2013
4. Deborah, H., Arymurthy, A.M.: Image enhancement and image restoration for old document image using genetic algorithm, published in 2010 Second International Conference on Advances in Computing, Control and Telecommunication Technologies (ACT), pp. 108–112, IEEE, 2010
5. Ueda, Y., Kuramoto, Y., Kubota, R., Suetake, N., Uchino, E.: An interactive genetic algorithm-based image sharpening system considering user's liking, published in IEEE Symposium on Computational Intelligence for Engineering Solutions (CIES), pp. 91–96, IEEE, 2013
6. Radlak, K., Smolka, B.: Visualization enhancement of segmented images using genetic algorithm, published in International Conference on Multimedia Computing and Systems (ICMCS), pp. 391–396, IEEE, 2014
7. Wu, Z.: Color Image Enhancement based on the rough set and adaptive Genetic Algorithm, published in International Conference on Mechatronic Science, Electric Engineering and Computer, Jilin, China, August 19–22, 2011
8. Dongzhou, M., Chao, X., Hongmei, G.: Hybrid genetic algorithm based image enhancement technology, published in International Conference on Internet Technology and Applications, pp. 1–4, IEEE, 2011
9. Munteanu, C., Rosa, A.: Evolutionary image enhancement with user behaviour modeling, published in Proceedings of the ACM symposium on Applied computing, pp. 316–320, ACM, 2001
10. Daniel, E., Anitha, J.: Optimum Green Plane Masking for the Contrast Enhancement of Retinal images using Enhanced Genetic Algorithm, published in Optik—International Journal for Light and Electron Optics,vol. 126, pp. 1726–1730, 2015
11. Hasikin, K., Isa, N.A.M.: Enhancement of the low contrast image using fuzzy set theory, published in 14th International Conference on Modelling and Simulation, pp. 371–376, IEEE, 2012
12. Chaira, T.: An improved medical image enhancement scheme using Type II fuzzy set, published. Appl. Soft Comput. **25**, 293–308 (2014)
13. Cepeda-Negrete, J., Sanchez-Yanez, R.E.: Automatic selection of color constancy algorithms for dark image enhancement by fuzzy rule-based reasoning. Appl. Soft Comput. **28**, 1–10 (2015)

14. Binaee, K., Hasanzadeh, R.P.R.: An ultrasound image enhancement method using local gradient based fuzzy similarity, published in Biomedical Signal Processing and Control, Vol. 13, pp. 89–101, 2014
15. Bing, Q., Lu, J., Jing, Z.: A Novel Image Enhancement Algorithm based on Information Fusion, published in International Conference on Computer Science and Software Engineering (IEEE), Vol. 1, pp. 577–580, 2008
16. Xiao-guang, Z., Ding, G., Jian-jian, X.: Generalized Fuzzy Enhancement of Image for Radiographic Testing Weld, published in Proceedings of the 4th International Symposium on Image and Signal Processing and Analysis, pp. 94–99, 2005
17. Balti, A., Sayadi, M., Fnaiech, F.: Segmentation and Enhancement of Fingerprint Images Using K-means, Fuzzy C-Mean algorithm and Statistical Features, published in International conference, pp. 1–5, 2011
18. Chaira, T.: Construction of Intuitionistic Fuzzy Contrast Enhanced Medical Images, published in Proceedings of 4th International Conference on Intelligent Human Computer Interaction, Kharagpur, India, pp. 1–5, December 27–29, IEEE, 2012
19. Zhang, D., Zhan, B., Yang, G., Hu, X.: An Improved Edge Detection Algorithm Based On Image Fuzzy Enhancement, published in IEEE, pp. 2412–2415, 2009
20. Wang, Y., Li, D., Xu, Y.: An Improved Image Enhancement Algorithm Based on Fuzzy Sets, published in IEEE, pp. 1–4,2013
21. Jiu, G.X., Jiao, J.F., Xiang, L.: Image Enhancement Method Based on Fuzzy Set and Subdivision, published in IEEE, pp. 174–176,2011
22. Wu, J., Yin, Z., Xiong, Y.: The Fast Multilevel Fuzzy Edge Detection of Blurry Images, published. IEEE Signal Process. Lett. **14**(5), 344–347 (2007)
23. Jia, W., Yang, J., Liu, Y., Fan, L., Ruan, O.: Improved Fast Image Enhancement Algorithm Based on Fuzzy Set Theory, published in Sixth International Conference on Intelligent Human-Machine Systems and Cybernetics, Vol. 2, pp. 173–175, 2014
24. Jinping, Z., Yongxiang, L., Linfu, D., Xueguang, Z., Jie, L.: A New Method of Fuzzy Edge Detection Based On Gauss Function, published in IEEE, Vol. 4, pp. 559–562, 2010
25. Jaya, V.L, Gopikakumari, R.: Fuzzy Rule based enhancement in the SMRT domain for low contrast images, published in Procedia Computer Science, Vol. 46, pp. 1747–1753, 2015
26. Saeed, F., George, K. M., Lu, H.: Image Enhancement using Fuzzy Set Theory, published in ACM, 1992
27. Rajua, G., Nair, M.S.: A fast and efficient color image enhancement method based on fuzzy-logic and histogram, published in International Journal Electronics Communication (AEU), Vol. 68, pp. 237–243, 2014
28. Hanmandlu, M., Jha, D., Sharma, R.: Color image enhancement by fuzzy intensification, published in Pattern Recognition Letters, Vol. 24, pp. 81–87, 2003
29. Alilou, V.K., Yaghmaee, F.: Application of GRNN Neural Network in Non-Texture Image Inpainting and Restoration, published. Pattern Recogn. Lett. **62**, 24–31 (2015)
30. Chitwong, S., Boonmee, T., Cheevasuvit, F.: Local Area Histogram Equalization based multispectral Image Enhancement from clustering using the competitive Hopfield neural network, published in CCGEI, Montrkal, Mayimai, IEEE, Vol. 3, pp. 1715–1718, 2003
31. Nieuwenhuis, C., Yan, M.: Knowledge based Image Enhancement using Neural network, published in the 18th International Conference on Pattern Recognition, Vol. 3, pp. 814–817, 2006
32. Yin, H., Liu, D.C.: Lateral Resolution Enhancement of Ultrasound Image using Neural Network, published in IEEE, pp. 1–4, 2009
33. Zhang, S., Lu, Y.: Image Resolution Enhancement using a Hopfield Neural Network, published in International Conference on Information Technology (ITNG'07), pp. 224–228, 2007
34. Pan, J., He, Y.: Recognition of plants by leaves digital image and neural network, published in International Conference on Computer Science and Software Engineering, Vol. 4, pp. 906–910, 2008

35. Singh, M., Singh, S.: Optimizing Image Enhancement for Screening Luggage at Airports, published in CIHSPS 2005—IEEE International Conference on Computational Intelligence for Homeland Security and Personal Safety Orlando, FL, USA, pp. 131–136, 31 March–1 April 2005
36. Ma, Y., Lin, D., Zhang, B., Xia, C.: A Novel Algorithm of Image Enhancement Based on Pulse Coupled Neural Network Time Matrix and Rough Set, published in Fourth International Conference on Fuzzy Systems and Knowledge Discovery (FSKD 2007), IEEE, Vol. 3, pp. 86–90, 2007
37. Rao, D.H.,: A Survey on Image Enhancement Techniques: Classical Spatial Filter, Neural Network, Cellular Neural Network, and Fuzzy Filter, published in IEEE, pp. 2821–2826, 2006
38. Weixin, G., Lianmin, S., Xiangyang, M., Nan, T., Xiaomeng, W.: X Ray Image Enhancement Technology for Steel Pipe Welding Based on Hopfield Neural Network, published in 2009 Second International Symposium on Computational Intelligence and Design, Vol. 2, pp. 107–110, 2009
39. Varghahan, B.Z., Amirani, M.C., Mihandoost, S.: Enhancement and Cleaning of handwritten Data by using Neural Networks and Threshold Technical, published in IEEE, pp. 1–4, 2011
40. Shanmugavadivu, P., Balasubramanian, K.: Particle swarm optimized multi-objective histogram equalization for image enhancement, published in Optics Laser Technology, Vol. 57, pp. 243–251, 2014
41. Draa, A., Bouaziz, A.: An artificial bee colony algorithm for image contrast enhancement, published in Swarm and Evolutionary Computation, Vol. 16, pp. 69–84, 2014
42. Gorai, A.,Ghosh, A.: Hue-Preserving Color Image Enhancement Using Particle Swarm Optimization, published in IEEE, pp. 563–568, 2011
43. Benala, T.R., Jampala, S.D., Villa, S.H., Konathala, B.: A novel approach to image edge enhancement using artificial bee colony optimization algorithm for hybridized smoothening filters, published in IEEE, pp. 1071–1076, 2009
44. Hanumantharaju, M.C., Aradhya, V.N.M., Ravishankar, M., Mamatha, A.: A Particle Swarm Optimization Method for Tuning the Parameters of Multiscale Retinex Based Color Image Enhancement, published in ICACCI'12, Chennai, T Nadu, India, ACM, pp. 721–727, August 3–5, 2012
45. Zhou, X., Sun, G., Zhao, D., Wang, Z., Gao, L., Wang, X., Jin, Y.: A Fuzzy Enhancement Method for Transmission Line Image Based on Genetic Algorithm, published in Ninth International Conference on Intelligent Information Hiding and Multimedia Signal Processing, pp. 223–226, 2013
46. Zhang, C., Lu, J.: Satellite Cloud Image Enhancement by Genetic Algorithm with Fuzzy Technique, published in International Conference on New Trends in Information and Service Science, pp. 1090–1095, 2009
47. Hoseini, P., Shayesteh, M.G.: Efficient contrast enhancement of images using hybrid ant colony optimization, genetic algorithm, and simulated annealing, published in Digital Signal Processing, Vol. 23, pp. 879–893, 2013

An Auto-Threshold Control for Isolating False Shadow from High-Resolution Images

K. Komathy

Abstract Digital image processing helps to extract the useful information from images for further investigations and analysis. Pre-processing of images, such as removal of noise and shadows, improves the quality of the high-resolution images for image recognition and tracking. In this paper, an optimal filter is recommended from its performance metrics for reducing speckle noise from the high-resolution images. The objective of the paper is to isolate the deceptive shadow from the original shadow, which is dark in color but is falsely treated as a shadow. Purging the false shadow is done by exploring the appropriate threshold value of a shadow region. Experiments show that our proposed method with auto-threshold control helps to isolate the deceptive shadow.

Keywords Noise removal · Shadow detection · Shadow removal · Filters · Speckle noise · Fake shadow

1 Introduction

The presence of shadow reduces the reliability of image processing algorithms while accounting for image tracking and analysis in fields such as medical, forensics, remote sensing, manufacturing, and defense, etc. Therefore, to improve the quality of output from those algorithms, the pre-processing plays a necessary stage. While segmenting, it uses the statistical features of the images, and the suspected shadows are extracted from the original image.

Dark objects may sometimes be included in the suspected shadows, and so a more accurate shadow detection method is needed to eliminate these dark objects

K. Komathy (✉)
Hindustan Institute of Technology and Science, Chennai, India
e-mail: kkomathy@hindustanuniv.ac.in

S. Urooj and J. Virmani (eds.), *Sensors and Image Processing*, Advances in Intelligent Systems and Computing 651, https://doi.org/10.1007/978-981-10-6614-6_27

from the shadow regions. Spatial information and spatial relationship between the objects of the image are generally used to trace the approximate positions of objects in the image, and this helps to demark the distinction of dark objects from the suspected shadow. In this paper, the objects dark in color, which could be mistaken for shadows, are ruled out via an automatic review of the threshold. For shadow removal, inner-outer outline profile line (IOOPL) [1] matching is used.

High resolution image, generally, is embedded with speckle noise due to electronic device interferences during the acquisition of images. Image interpretation and recognition are very difficult with this type of noise. Filters are applied to remove noises completely or partially, which enhance and improve the quality of an image. Different types of filters [2] that can be used to remove unsolicited pixels as noise from the image are linear filtering, median filtering, and adaptive filtering. The performance of these filters is also examined in this paper so as to find out the most suitable filter for removing the speckle noise from the high-resolution images.

2 Related Works

K. L. Chung et al. [3] presented a novel threshold scheme to detect shadows more accurately by stretching the gap between the ratio values of shadow and non-shadow pixels. H. Ma et al. [4] used a normalized saturation-value difference index (NSVDI) in hue-saturation-value (HSV) color space to trace shadows and recovered the information under shadows using histogram matching. Lorenzi et al. [5] reconstructed the image using linear regression where the intensities of the shaded pixels are tuned using that of non-shadow regions. P. Sarabandi et al. [6] proposed a nonlinear transformation for separating the shadow from dark objects present in the image. Turlapaty et al. [7] proposed a hybrid approach, where the feature, namely *relative building height,* is extracted from images, to separate the buildings from non-buildings.

Qingxiong Yang et al. [8] used the bilateral filtering for shadow removal from single input image. The luminance contrast in regions with similar surface reflectance is reduced, while the contrast in regions with different surface reflectance is preserved. Wen Liu et al. [9] proposed an object-based classification method that compares the brightness of the region of interest with the neighboring area. Then, a correction is applied using a linear function to produce a shadow-free image. Hongya Zhang et al. [1] discussed Otsu's convexity model. However, no image calibration is done for the intensity and the illumination adjustments. An attempt is made to discern the false shadow from the original shadow to enhance the quality of the image based on the radiometric profile of each shadow region. Figure 1 shows a few sample images from the dataset that includes 500 images of high-resolution [10] generated by Herve Jegou et al. [11] for their work on large-scale image search. Figure 2 lists the processes involved in shadow detection and removal.

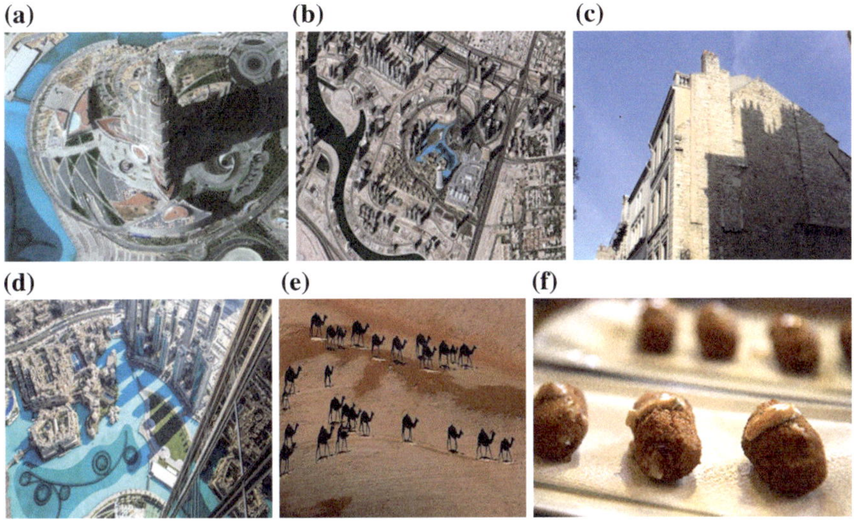

Fig. 1 A Sample of High Resolution Images with Shadow

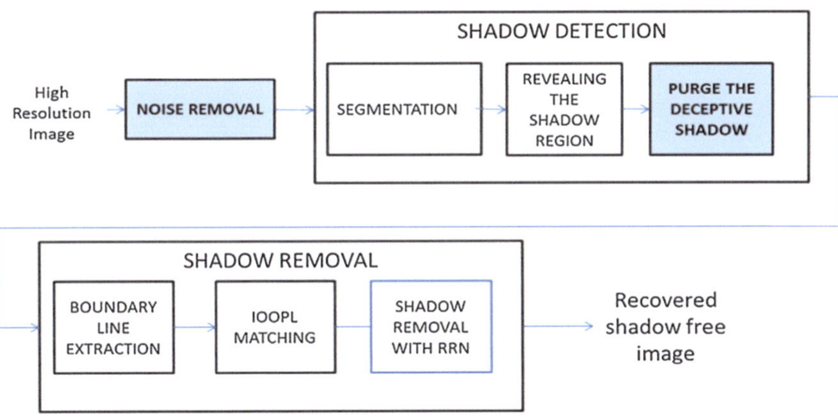

Fig. 2 Process Flow for Shadow Removal

3 De-Noising

De-noising of high-resolution images using linear filtering, median filtering, and adaptive filtering is studied and investigated. Gaussian filter [12], median filter [13], average filter [14], and Weiner filter [15] are compared to find out the best filter to remove the speckle noise present in the high-resolution images.

The performance of the filters [14] is compared based on the metrics such as peak signal-to-noise ratio (PSNR), mean square error (MSE), and average

Table 1 Quality metrics of Gaussian and median filters for speckle noise

Sample image	Performance of filters					
	Gaussian			Median		
	PSNR	MSE	AD	PSNR	MSE	AD
Image (a)	35.51	8.282	0.0664	28.02	102.414	−0.0434
Image (b)	32.42	37.204	0.0466	24.996	205.796	0.1480
Image (c)	44.52	2.292	0.0229	37.13	12.589	−0.0037
Image (d)	47.36	1.1927	0.0160	41.55	4.546	−0.0555
Image (e)	35.25	19.404	0.0181	28.40	93.920	0.0826
Image (f)	34.37	23.754	0.0873	26.63	141.209	0.2184

Table 2 Quality metrics of average and Wiener filters for speckle noise

Sample image	Performance of filters					
	Average			Wiener		
	PSNR	MSE	AD	PSNR	MSE	AD
Image (a)	26.86	133.710	0.2089	31.33	47.770	0.0482
Image (b)	23.82	269.311	0.1451	27.75	108.995	−0.0043
Image (c)	35.45	18.496	0.0738	38.70	8.757	−0.0035
Image (d)	38.30	9.609	0.0516	43.07	3.204	0.0154
Image (e)	26.60	141.95	0.0566	31.26	48.564	0.0020
Image (f)	25.91	166.612	0.2707	30.39	59.388	−0.0031

difference (AD). An optimal filter shows a high PSNR value, a lower MSE value and a high AD value. Performance of filters is recorded for the sample images shown in Fig. 1 as given in Table 1 and 2. The tables compare the quality measures such as PSNR, MSE, and AD for Gaussian, median, average, and Wiener filters when applied for speckle noise and it is found that Gaussian filter gives a better performance when compared to other filters.

4 Shadow Detection

Shadow detection process discovers the shadow region from the image. High resolution image contains more affluent spatial information. In order to use spatial information to detect shadows, image segmentation is needed. Traditional methods for segmenting images [4] are likely to result in inadequate segmentation, which makes it difficult to separate shadows from dark objects. The combined morphological and the convexity model are used for segmenting the shadow region [16]. Let CM is the convexity model, where

$CM(R_i)$ is defined in the R_i on predicate logic,
$F(R_i)$ is the average features vector,
$S(R_{i,} R_m)$ is the similarity of neighbor region,
E is the entropy of morphology,
R_i is the area, and
BM(E) represents the morphological operations on binary image.

If R_i obeys the convex model such that $CM(R_{i)}$ = TRUE, then the average feature vector $F(R_i)$ will be more or less than all the neighbors around the region.

$$S\left(R_{i,} R_m\right) < S(R_m, R_n); R_m, R_n \in N_i, \; m \neq n \tag{1}$$

$$E = \mathrm{ENTROPY}(I), \quad \text{for } I = 0, 1, \ldots, 255 \tag{2}$$

where E represents the entropy of intensity of the image. Entropy is a statistical measure of randomness that can be used to characterize the feature of the input image. Entropy value is obtained from the expression (–sum (p.*$\log_2$(p)) where p represents the histogram counts.

$$S = f(CM) \tag{3}$$

where S is the output value of the center pixel in the mxn block. Morphological general sliding neighborhood operations call convexity model function f(CM) for each pixel in the binary image for segmentation. Zero padding will be done for the mxn block at the edges, if necessary. Figure 3 shows the original image and the output of the convexity and morphological segmentation method.

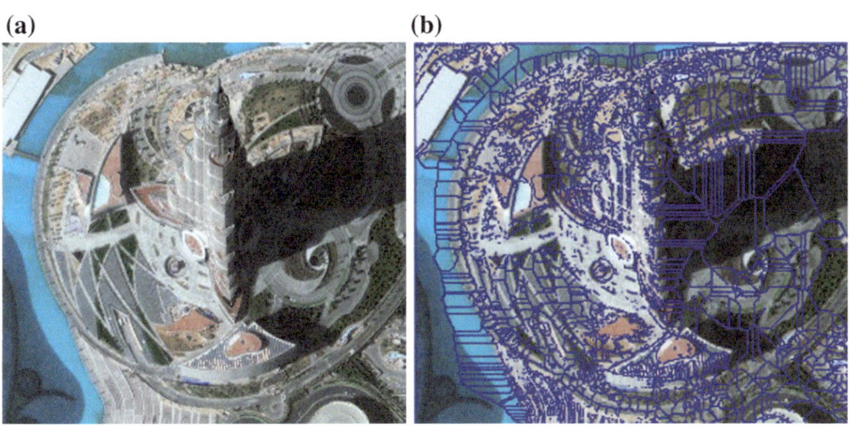

Fig. 3 **a** Original Image **b** Image after Convexity and Morphological segmentation

To avoid misclassification, a proper threshold can detach shadow from the non-shadow region. A set of threshold values can be mapped from the histogram of the original image, and then, it is revealed the shadow objects by comparing the grayscale average and the threshold of each object obtained after segmentation. The grayscale value is chosen as shown in the Eqs. (4) and (5).

$$\mathrm{Gq} = (\mathrm{Ga} + \mathrm{Gs}) \tag{4}$$

$$\mathrm{h(T)} = \min\{\mathrm{h(Gq} - \varepsilon),\ \mathrm{h(Gq} + \varepsilon)\} \tag{5}$$

where Ga is the average grayscale value of an image; Gs is the left peak of the shadow in the histogram; T is the threshold; ε represents the neighborhood of T, where $T \in [Gq - \varepsilon, Gq + \varepsilon]$; and h(I) is the frequency of I, where I = 0, 1, … , 255.

In shadow and non-shadow regions, the grayscale difference at the red and green wavebands is more noticeable than at the blue waveband. A shadow can be reclaimed using the threshold method considering the red and green wavebands. Explicitly, an object is determined to be a shadow if its threshold is more than the grayscale average in both red and green wavebands [13]. Figure 4a shows the resultant image with shadow identified.

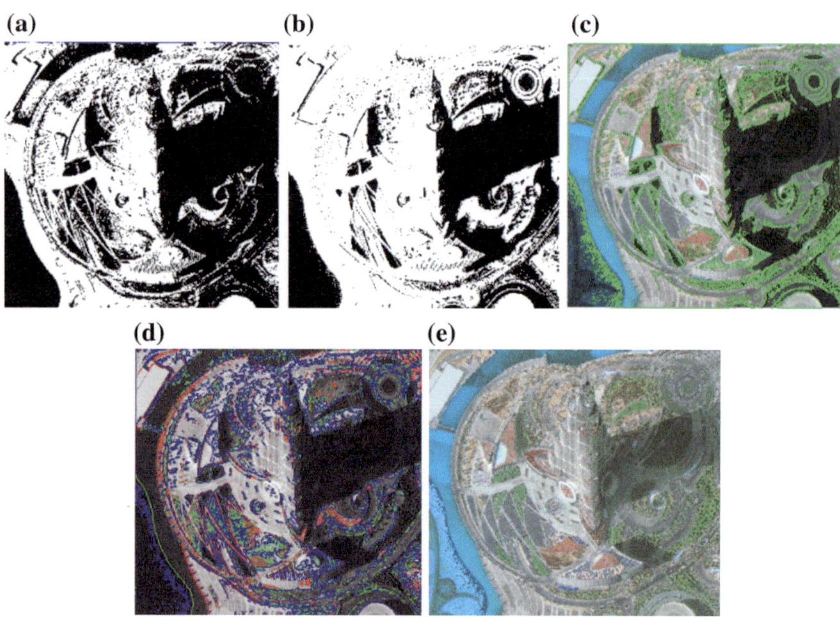

Fig. 4 Resultant image after **a** detecting the shadow **b** purging the shadow **c** tracing the boundary of the shadow region **d** IOOPL boundary matching **e** RRN and PF methods

4.1 Purging the Deceptive Shadow

Shadow purging is the process of eliminating the deceptive shadow from the image. Threshold-based method is recommended to eliminate the deceptive shadow. While extracting the features such as grayscale value of an image, the threshold value is also added to isolate the shadow region from the original image. The grayscale values of a shadow area and a non-shadow area in the blue (B), the red (R), and green (G) wavebands are obtained by applying Rayleigh scattering method. It is found that the difference is smaller in the B than in R and G bands. Following is the algorithm that gives the steps involved in automatic threshold calculation.

Algorithm: Purging the deceptive shadow

Pseudo code:

Step 1: Convert grayscale image into a binary image
Step 2: Calculate the grayscale threshold value of the image
Step 3: Fix the threshold value as the grayscale value of red channel
Step 4: If the average grayscale value of the image exceeds the threshold value it will be considered as part of the shadow region
Step 5: False shadow regions are ruled out by isolating the threshold of blue channel from threshold of green channel
Step 6: If the average grayscale value falls below the threshold value, it will be considered as a false shadow region. Then the shadow-free image is obtained by purging the deceptive shadow.

Table 3 shows the threshold values that are arrived for gray scale, green, red, blue channels, and false shadow for six sample images given in Fig. 1. Figure 4b displays the resultant image after purging the shadow. Time complexity of the purging algorithm is found to be O(n), where n is the size of the image in terms of nxn pixels.

Table 3 Threshold value to fragment shadow from image

Images	Gray-scale	Red channel	Green channel	Blue channel	False shadow
Image (a)	0.4471	0.4078	0.4705	0.4431	0.0274
Image (b)	0.4275	0.4313	0.4916	0.4078	0.0838
Image (c)	0.6510	0.6745	0.6392	0.6313	0.0079
Image (d)	0.4745	0.5960	0.4745	0.4862	0.0117
Image (e)	0.3686	0.4313	0.3803	0.3960	0.0157
Image (f)	0.4627	0.5921	0.4235	0.3215	0.1020

5 Shadow Removal

5.1 Shadow Boundary Line Extraction

Boundary of the binary image shadow region is achieved by Moore-Neighbor Tracing Method [17]. Then, the shadow region of the binary image is mapped to the original image by contour tracing, which plots the exact boundary of a shadow region in original image as shown Fig. 4c.

5.2 IOOPL Boundary Matching

IOOPL [1] works on the inner and outer layer extraction from the image. The inner depicts the red channel from the image; the middle region refers to the green channel; and the outer layer represents the blue channel [6]. IOOPL matching is a process of finding similarity matching between sections to identify the homogeneous sections. Figure 4d shows the output after IOOPL.

5.3 Shadow Removal Using RRN and PF

Shadows are then removed by using the homogeneous sections obtained by line pair matching. It follows two processes for removing the shadow. First approach called *relative radiometric normalization* (RRN) [18] calculates the radiation parameter according to the homogeneous points of each object, and then applies the relative radiation correction to each object. The second approach collects and analyzes all the homogeneous sections for polynomial fitting (PF) [19] and retrieves all shadows directly with the obtained fitting parameters.

6 Conclusion

Investigation is done for finding an optimal filter to remove the speckle noise from high-resolution images. Gaussian filter is found to reduce the speckle noise efficiently. Then, the proposed auto-threshold method is able to recover the darker objects of the images by ignoring the deceptive shadows adequately from the images. Most of the shadows are ruled out, and hidden objects are clearly visible up to 85%. Our proposed method is based on the spatial relationships between the objects of an image, and false shadows are effectively detected. IOOPL matching method is used to restore the information in the shadow area. The radiation difference estimated between inner and outer lines of homogeneous sections of an object has helped to recover the shadow effectively. Further, enhancement of the proposed method is to provide 100% shadow removal.

References

1. Zhang, H., Sun, K., Li, W.: Object-oriented shadow detection and removal from urban high-resolution remote sensing images. IEEE Trans. Geosci. Remote Sens. **52**(11), 6972–6982 (2014)
2. Pathak, A., Singh, E.: Comparative study on filtering techniques of digital image processing. J. Adv. Electr. Comp. Eng. **4**(6), 669–674 (2014)
3. Chung, K.L., Lin, Y.R., Huang, Y.-H.: Efficient shadow detection of color aerial images based on successive thresholding scheme. IEEETrans. Geosci. Remote Sens. **47**(2), 671–682 (2009)
4. Ma, H., Qin, Q., Shen, X.: Shadow segmentation and compensation in high resolution satellite images. IEEE IGARSS **2**, 1036–1039 (2008)
5. Luca, L.: Innovative methods for the reconstruction of new generation satellite remote sensing images. Doctoral Thesis. (2012)
6. Sarabandi, P., Yamazaki, F., Matsuoka, M., Kiremidjian, A.: Shadow detection and radiometric restoration in satellite high resolution images. IEEE IGARSS, Sep. **6**, 3744–3747 (2004)
7. Turlapaty, A., Gokaraju, B., Qian, D., Nicolas, H.Y., Aanstoos, V.A.: A hybrid approach for building extraction from spaceborne multi-angular optical imagery. IEEE Tr Sel. Topics Appl. Earth Observ. Remote Sens. **5**(1), 89–100 (2012)
8. Yang, Q., Tan, K.-H., Ahuja, N.H.: Shadow removal using bilateral filtering. IEEE Tran. Image Processing **21**(10), 4361–4368 (2012)
9. Wen Liu:Object-Based Shadow Extraction and Correction of High-Resolution Optical Satellite Images. IEEE J. Sel. Topics Appl. Earth Observ Remote Sens. **5**(4), 1296–1302 (2012)
10. The INRIA Holidays dataset. Downloaded from http://lear.inrialpes.fr/~jegou/data.php
11. Herve, J., Matthijs, D., Cordelia, S.: Hamming embedding and weak geometry consistency for large scale image search. In: 10th European conference on Computer vision. pp. 314–324. Marseille, France, (2008)
12. Jalobeanu, L., Féraud, B., Zerubia, J.: An adaptive Gaussian model for satellite image deblurring. IEEE Trans. Image Process. **13**(4), 613–621 (2004)
13. T. Liao, G., Gallagher, N.: Statistical analysis of two dimensional median filtered images. In: IEEE International Conference Acoustics, Speech, and Signal Processing. 9, pp. 255–258 (1984)
14. Eskicioglu, M., Fisher, P.S.: Image quality measures and their performance. IEEE Trans. Commun. **43**, 2959–2965 (1995)
15. Yoo, J.-C.: Chang Wook Ahn: Image restoration by blind-Wiener filter. Image Processing, IET **8**(12), 815–823 (2014)
16. Sin, L.L., Daya Sagar, B.S., Voon, C.K., Lea T.T: Morphological convexity measures for terrestrial basins derived from digital elevation models. Comput. Geosci. **37**, 1285–1294 (2011)
17. Toussaint, G.: Course Notes: Grids, connectivity and contour Tracing. Downloaded from the Website. http://jeff.cs.mcgill.ca/~godfried/teaching/pr-notes/contour.ps
18. Mateos, C.J.B., Carlos Pinilla, Ruiz, ACS, Rubén González Crespo: Relative Radiometric Normalization of Multitemporal images. Int. J. Artif. Intell. Interact. Multime. **1**(3), 54–59 (2010)
19. Ameer, S.: Investigating Polynomial Fitting Schemes for Image Compression. Thesis, Waterloo, Ontario, Canada (2009)

MIX
Papier aus verantwortungsvollen Quellen
Paper from responsible sources
FSC® C105338

If you have any concerns about our products,
you can contact us on
ProductSafety@springernature.com

In case Publisher is established outside the EU,
the EU authorized representative is:
Springer Nature Customer Service Center GmbH
Europaplatz 3, 69115 Heidelberg, Germany

Printed by Libri Plureos GmbH
in Hamburg, Germany